Simulation
in Manufacturing

C^3 INDUSTRIAL CONTROL, COMPUTERS AND COMMUNICATIONS SERIES

Series Editor: **Professor Derek R. Wilson**
University of Westminster, England

Simulation in Manufacturing

Norman Thomson
Strathclyde University, UK, formerly IBM(UK)

RESEARCH STUDIES PRESS LTD.
Taunton, Somerset, England

JOHN WILEY & SONS INC.
New York · Chichester · Toronto · Brisbane · Singapore

RESEARCH STUDIES PRESS LTD.
24 Belvedere Road, Taunton, Somerset, England TA1 1HD

Marketing and Distribution:

Australia and New Zealand:
Jacaranda Wiley Ltd.
GPO Box 859, Brisbane, Queensland 4001, Australia

Canada:
JOHN WILEY & SONS CANADA LIMITED
22 Worcester Road, Rexdale, Ontario, Canada

Europe, Africa, Middle East and Japan:
JOHN WILEY & SONS LIMITED
Baffins Lane, Chichester, West Sussex, England

North and South America:
JOHN WILEY & SONS INC.
605 Third Avenue, New York, NY 10158, USA

South East Asia:
JOHN WILEY & SONS (SEA) PTE LTD.
37 Jalan Pemimpin 05-04
Block B Union Industrial Building, Singapore 2057

Library of Congress Cataloging-in-Publication Data

Available

British Library Cataloguing in Publication Data

A catalogue record for this book
is available from the British Library.

ISBN 0 86380 172 2 (Research Studies Press Ltd.) *[Identifies the book for orders except in America.]*
ISBN 0 471 95738 0 (John Wiley & Sons Inc.) *[Identifies the book for orders in USA.]*

Printed in Great Britain by SRP Ltd., Exeter

Contents

Figures

EDITORIAL PREFACE

Microprocessors are now advertised on Prime Time TV along with various brands of soft drinks and lagers! Such is the rapid integration of Information Technology into the domestic life cycle. That well known TV personality, Clive James, frequently refers to the 'Information Highway' as he conducts on-line interviews across the world directly into every home. It is expected that Video on Demand, Banking, Shopping and an even greater range of Information Services will be available in the home by the end of the decade, through the advent of fibre into the home.

To support this explosion in Information facilities, Manufacturing has now progressed from an ART into an exact science. It is no longer, 'Can it be done?' - but rather, 'What is the optimum way to do it?'; and for that, Simulation is an essential evaluation tool.

Simulation technology is now the ubiquitous tool of all manufacturing industries. Global competition is such that business survival is dependent upon that simple idea: 'getting it right first time'. For that, Simulation is essential, because it is necessary to explore all the possible options before committing resources to the production process.

Norman Thomson has spent 20 years simulating and evaluating manufacturing strategies for the IBM Corporation. During this time he has used and evaluated a significant range of simulation tools. For this book he has chosen the software tool 'AIM' to illustrate the generic concepts that sustain and support manufacturing.

It is a book that will be extraordinarily valuable to students of manufacturing, who need to understand the principles of manufacturing simulation; and it will be of particular benefit to practising engineers, who have to optimise the production cycles in today's globally competitive environment. The key ingredient of Norman Thomson's book is that his presentation is based on the fact that he has been practising his profession and 'doing it for real' in the highly competitive world of manufacturing computer products.

Production of computer products has proved to be an industry in which the application of advanced manufacturing techniques is absolutely essential to business efficiency. Norman Thomson has alway been at the forefront of manufacturing simulation technology. He is a world-class expert; and it is a great pleasure to welcome this volume to the Series, so that his experience, expertise and advice are available to students and practising engineers.

Derek Wilson
London 1995

PREFACE

This is a book about how to do manufacturing simulation in a windows computing environment. Simulation systems are an evolving form of computer package. In 1985 Professor Alan Carrie of Strathclyde University wrote that there was a requirement for simulation systems which spoke to their users in manufacturing rather than computing terms. That gap has now been filled and there is at least one simulation system called AIM which fulfils this goal in an extremely comprehensive manner. Windows mean that communication is no longer by means of text files containing computer programs which have to be compiled or interpreted. Instead of writing programs, model-building is now an interactive process involving mouse clicks, list box selections, radio buttons and similar paraphernalia.

This book has the overriding objective of portraying the various facets of simulation modelling as a live practical experience in this sort of environment. It describes a manufacturing vocabulary with which to talk to the simulation system. The core activity of simulation is still the hard slog of building models, but now the primary input operation is windows activity rather than copy-typing. Thus instead of program code which was characteristic of earlier books on simulation the reader will find instructions on how to interact with AIM.

Many books on discrete event simulation are devoted to instructing the reader in detail on how to use a single language or package. Ideally this book would be written in a system-independent way; however, realism demands that *some* system has to be used, just as in the preceding computer age authors had to choose one particular computer language such as GPSS or SLAM. Other books on simulation go deeply into the associated mathematics and statistics. The practising simulation analyst should indeed delve from time to time into these more theoretical aspects. The approach taken in this book is to try and portray the balance between these various strands of simulation modelling. AIM is chosen as the simulation system, and the list of references is confined to what might be judged a useful set of reference books which are desirable to have to hand when deeper cover of one of the ancillary topics is needed.

INTRODUCTION

The foundation of this book is the description of a modelling framework and vocabulary for constructing discrete event simulation models in manufacturing. The framework is independent of any particular simulation package, and therefore is of general value for describing a wide range of manufacturing processes at a conceptual level. It could for example be used in conjunction with programming ideas for constructing a simulation package from scratch in a suitable high-level language, as is described in Ref. [3]. However, as computers become more sophisticated, simulation practitioners are increasingly moving away from this algorithmically based approach, and demand instead computer packages which act as interactive model-builders and talk to their users in manufacturing rather than computing terms. They should also have simple but good graphics with the potential for linking to CAD systems to provide engineering drawings as backgrounds, and the ability to share dynamically linkable data in engineering and other databases. As far as algorithms are concerned, the number of generic rules for managing and operating manufacturing systems is not inordinately large, and a good package should contain a repertoire of these which is adequate to deal with the vast majority of situations encountered in practice. Any possible gap should be covered by the provision of a link to user-written routines in a general-purpose high-level programming language. All of the above require that users accept a discipline in which terms are used consistently and with the sort of precision that good software design demands. Hence the modelling philosophy this book seeks to convey.

Attempts to impose common standards of manufacturing terminology on a world-wide basis are not new - see for example the Reference Model for Shop Floor Standards published by ANSI/NEMA (American National Standards Institute/ National Electrical Manufacturers' Association) in 1988. It is doubtful, however, whether any such descriptive system will achieve much general acceptance without the motivation of a computer package which implements its constructs and provides meaningful realisations of models constructed using it. Such a system is AIM*, manufactured by the Pritsker Corporation of Indiana in association with IBM as Business Partners. This product, unlike previous simulation packages and languages, has established a *de facto* vocabulary which is of value wherever any analysis of manufacturing processes is being carried out whether for simulation or for any other decision-making or analytic purpose.

The usage of earlier major simulation languages such as GPSS, SLAM and SIMAN was characterised by devices and programming "tricks" whereby the realities of the manufacturing world required to be ingeniously matched to the constructs of the programming language. Modern expectation is for a simulation system which incorporates once and for all into a

---

* AIM is distributed in the US by Pritsker Corporation, Suite 500, 8910 Purdue Road, Indianapolis IN 4628-1170, and in the UK by Computer Sciences Ltd., Guild Centre, Lords Walk, Preston, Lancashire PR1 1RE.

higher level package all the necessary programming skill and ingenuity which might be called upon for manufacturing modelling. Programming skill is thereby replaced with the requirement that the user possess a focussed understanding of what is common to *all* manufacturing processes regardless of the objects being manufactured. To describe the vocabulary of AIM as a philosophy of manufacturing is over-grandiose, but it does provide a basis for thinking about manufacturing processes at a first level of abstraction using terms which should be clearly understood and unambiguous. Further, it allows its users to talk to the computer in manufacturing rather than computing terms, which is a major step forward for manufacturing simulation systems. It also establishes a specification language which should be of value when simulation is undertaken in other ways, for example by writing more limited programs or libraries in a general purpose programming language. In this sense computers are acting as agents for standardisation.

Anyone performing simulation in manufacturing, at whatever level of detail, and regardless of what computing system they are using, must go through much the same industry-independent analysis. IBM's alliance with the Pritsker Corporation came about as the result of a very comprehensive requirements analysis for simulation conducted by IBM personnel, both within IBM plants and in the course of visits to customers across a wide spectrum of industries. The conclusion of this was a decision that AIM, which was by that time already well developed, promised, with appropriate adaptations, to fulfil closely the perceived world-wide needs. This book endeavours to capture those aspects of the outcome of that work which are of general significance, regardless of whether or not the reader is fortunate enough to have access to an AIM system.

Discussion of general terms and concepts is intermingled with instructions for using the version and release of AIM which was available at the time of writing. It seemed worth while to give AIM instructions in some detail even to the extent in the early chapters of giving precise GUI (Graphic User Interface) actions needed. This is the analogue of giving full program listings plus compilation etc. instructions in the pre-windows era. The reader who does not have access to an AIM system should nevertheless be able to read the boxes in which AIM instructions are enclosed and see the broad pattern of those things which have to be done whatever system is employed. It was a conscious decision not to include images of the various GUI user-interface screens since these may change as the product evolves, whereas the broad sequence of user-driven GUI events is much less likely to do so. In summary, this book describes a terminology for specifying and conceptualising manufacturing processes with a view to their simulation. Since the designers of AIM did more or less that same thing it seemed worth while to use the opportunity to provide simultaneously a guide which might accelerate the progress of newcomers to the AIM package.

It is hoped that the book will prove useful to two sorts of people, first, students in the disciplines of Industrial and Manufacturing Engineering, and secondly, practitioners in industry itself. For the former, simulation should be an important ingredient in their courses but cannot in itself form more than a small part of the total picture. On the other hand those with practical experience of constructing simulation models in industry know that an individual model of any substance comes about only as the result of a long, laborious and painstaking process. The requirement of courses must thus be to present the flavour of such activities while short-circuiting the more humdrum and time-consuming phases of development. A course on simulation should also impress on students that it is an essentially *practical* activity; at the time of writing there are few texts with simulation in their title which emphasise this point.

This may also be a matter for concern for the practitioners in industry who make up the remainder of the intended readership. There are numerous and excellent books which stress the theoretical and statistical aspects of simulation but understate the actualities of producing models in a real industrial environment. This is an activity which is often devolved in practice to engineers, scientists or other skilled people whose primary disciplines are not Operational Research or statistics. It is one of the aims of this volume to provide such users with an account in compact form of the fundamental information and techniques which they should use and modes of thinking which they should adopt in order for them to start modelling effectively. If it happens also that they have the opportunity to use AIM as their simulation package, then the level of detail will be that appropriate to an extended tutorial introduction which should take them rapidly to the point where they get "under the skin" of the product and are ready to start building their own models with confidence.

The authors of Ref. [3] explain in their preface how they came to write their book because they had failed to find a text which explained concisely how to develop and experiment with a discrete-event simulation model represented as a computer program. Now five years later, that is, an aeon away in terms of the development of the use of personal computers, this book is written because the analogous text on how to use modern simulation systems appears not to exist. The aim has been not only to record in succinct terms what is important, but also to do this in a way which reflects the balance of importance of the various components. The simulation system, although central, is just one part of the whole. Other adjuncts are a statistical computing package, some skill in statistics and some knowledge of Queueing Theory and Numerical Analysis. For example, when estimates of operation times and inter-arrival times are required some understanding of statistical distributions is necessary, together with the computational power to estimate their parameters. These are discussed in Chapter 3 in a fashion which it is hoped will make them less fearsome than some of the texts on this subject might have their readers suppose. The number of books which describe distributions is vast. Ref. [4] is devoted entirely to distributions while in Refs. [2] and [7] they form a small but valuable part of the subject matter. When a point is reached in modelling where a distribution has to be fitted, a statistical package is a necessary adjunct to a simulation package. Again the choice is large, and the distribution-fitting routines would probably form only a small part of the statistics package.

Another consideration is that the results of simulation should stand up to numerical common sense. Sometimes this need involve no more than rapid checks with a calculator. Where the distributions concerned have simple and well known properties, analysis derived from Queueing Theory as opposed to simulation can provide formulae which give rapid estimates and checks. The most frequently used of these are described in Chapter 6. For those who need or would like to go beyond this point two books on Queueing Theory Refs. [1] and [10] are recommended as very readable texts.

At a later stage a project may come to the point of experimentation (see Chapter 8) at which point some knowledge of Design and Analysis of Experiments is often a requirement if the fullest use is to be made of a carefully and laboriously crafted model. Ref. [9] is recommended for its modern approach to Design of Experiments. At this stage another sort of computer package (or perhaps another component of the previous one) is needed, and again the choice is vast.

Throughout model development and deployment it is necessary for the user to involve, consult and generally win the confidence of the people whose day to day activities he or she is modelling. This is a matter more of psychology and personality than of science or statistics, and it is not in general the sort of skill which is acquired from books. It is nevertheless of great importance in the art of making simulation effective.

This book then is intended to be an approachable and practical guide to simulation in manufacturing based on years of doing just this, while at the same time taking account of the ways in which the wider progress of computing has influenced the development of simulation packages, particularly on personal computers and work stations. It attempts to portray in its own balance of material the sort of balance which exists in practice between the various components and subskills involved in making persuasive and worthwhile models in the context of manufacturing. It may seem at first sight that there is heavy emphasis on the pedestrian details of communicating with the AIM computer package. This is deliberately so, to emphasise that this is what real simulation is like, that is, long stretches of relatively patient and dull initial construction before the heights are scaled and the more exciting phases of analysis and results begin to come into view.

It is hoped that the students and practitioners to whom it is addressed will find it to be of some service.

1. BASIC MANUFACTURING CONCEPTS

1.1 A Brief History of Simulation in Manufacturing

It is not hard to conceive that if a computer can be programmed to execute the steps of a mathematical algorithm, which is in essence a transformation of numbers, characters, etc. organised into datatypes, it can also be directed to mimic transformations and movements of material in the real world and in particular the real world of manufacturing. Around the late 1950's it became apparent that the high level languages of that time, Fortran, Cobol and the like, were very inadequate for this second type of programming, and another type of language was invented - the simulation language - of which GPSS, developed by IBM, was the first major example. Later languages possessed the same broad characteristics, notably SLAM, SIMAN and SIMSCRIPT, and also the general purpose language SIMULA whose list-oriented approach made it appropriate for use in simulation. (It

1

seemed to be a fashion of these times that simulation languages should have names beginning with S!)

Simulation has always been a highly CPU-intensive form of programming, and in its early days could not have been contemplated on any other than relatively large mainframe computers. When personal computers came along in the early 1980's a new possibility became available, namely that of having the user interact directly with models in the course of execution, and this capacity for so-called *Visually Interactive Simulation* often more than compensated for the relative slowness of the small stand-alone machine. SEEWHY from British Leyland was a pioneering system in these days and this has emerged through several changes of ownership and revisions to become what is now known as WITNESS.* Many similar but less extensive simulation systems have subsequently come to (and in many cases gone from) the market. A complete catalogue of such systems that appeared in the 1980's would take a long time to compile. This has not been attempted since what is important today is the parallel evolution of windows systems with consequent changes in patterns of computing behaviour across an ever-increasingly computer literate population at large.

It is against this background that IBM set out in the late 1980's to research the requirements for a modern simulation system, and chose as Business Partners the Pritsker Corporation whose existing product AIM was deemed to be readily adaptable to these goals. Although AIM is still the subject of ongoing development, its contribution to the history of simulation

---

* WITNESS is developed and marketed by AT&T Istel, Highfield House, Headless Cross Drive, Redditch, Worcs B97 5EQ (UK), and 25800 Science Park Drive, Suite 100, Beachwood, Cleveland OH44122 (US).

languages is assured as the first major system to talk to its users entirely in manufacturing terms.

1.2 General Requirements for Simulation Software

One view of a simulation system within the broad compass of a manufacturing complex is that of a stepping stone on the way to implementing a manufacturing control system. A major problem which companies have in taking this step is the multiple skills and experience base required of the simulation analyst. As hinted in the introduction, the ideal analyst possesses a mixture of mathematics, statistics, computer science and engineering, and it is hard for even large companies to maintain these composite skills when the introduction of new manufacturing lines and their associated control systems is at best intermittent.

A first step to alleviate this problem would be to have a unified view of manufacturing control systems through all the phases of their life cycle, that is, design, analysis, implementation, operation, maintenance and modification. At all of these stages simulation is appropriate, and hence it is desirable that the unified view of control systems extend seamlessly to a simulation system. The vision is one in which an entire factory computer system, including operational software and data collection systems, is available to facilitate simulation in the most convenient manner. The principal functions to be found universally in factories are:

Engineering Management;
Production and Business Planning;
Plant Operation;
Finance and Accounting.

3

Simulation might use data produced by any combination of these functions. Some examples of questions it might be called on to resolve are:

An industrial engineer wants to know if the purchase of an additional machine will alleviate a bottleneck operation known to be impeding throughput on the shop floor.

A market planner wants to know if the factory can deliver a major new order without missing existing schedules.

A process engineer is about to release a process change to reduce the cycle time for a product, and wants to evaluate how it will work with next month's mix of products.

The production control supervisor needs to improve the scheduling of a flexible manufacturing line to increase throughput, and wants to know if there is a scheduling algorithm that will perform better than the current one.

A line manager wants to work out an optimum strategy when a particular machine breaks down - is it better in the long run to switch production to another line or to await the repair to the first.

The production planning manager wants to know whether it is cost effective to transfer some parts of the work of the factory to a vendor.

1.3 Vocabulary and Typography

In this chapter, and indeed through this entire book, basic manufacturing concepts are described in a generic vocabulary which leans heavily on that to be found in AIM. This in turn was influenced by what IBM perceived as a good basis for taking a unified approach to viewing manufacturing control systems.

Items which have specific meanings in the context of the book are spelt with an initial capital letter. On their first occurrence they are additionally printed in **bold** type. Some parts of the generic vocabulary are introduced in the form of direct instructions to users of AIM. These appear in boxes, within which bold type is used to indicate words on the screen to which the user's eye is directed.

1.4 Projects and Alternatives

The major unit recognised by the analyst of industrial processes is the **Project**. Each Project has its own **Database**, within which are one or more **Alternatives**, that is, models which reflect different options either in terms of structure or parameters. Alternatives share a common **Project Description** (that is, a short character string description) and a common **Problem Configuration**. A Problem Configuration is a list of output performance measures which can be adjusted according to the overall purpose of the enquiry underlying the simulation.

The term Alternative will be used in contexts where the reader might expect to find the word "model." Model and Alternative may be taken as

5

synonyms, although the latter will be almost always used in what follows. It may seem strange to use the word Alternative rather than what might seem to be the more expressive and simpler term "model." However "Alternative" more accurately reflects an essential attribute of most models (as the word is used in the present context) in that their underlying purpose is to mimic existing or planned configurations of people and materials, and investigate *alternative* ways of deploying these without having to experiment with the real life objects.

An Alternative is a static entity like a computer program. A single execution of an Alternative is a dynamic entity analogous to a single run of a computer program. When such an execution is performed within the context of an experiment it is called a **Replicate**. A general simulation experiment thus consists of several Replicates, the differences between which reflect the varying outputs from different random number streams which are invoked within the Alternative.

Part is a fundamental concept. It is used to denote a primary *type* of object manufactured. In different contexts this might be a printed circuit board of one specific design, a motor car of one particular model and set of options, or a tub of margarine of a given recipe and volume. A Part is thus characterised by possessing a unique Bill of Materials. The term Part is also used less exactly to denote an *instance* of such a type, for example in talking about a throughput of a thousand Parts per shift. In this sense a Part could be defined as a primary unit of production. In practice, few problems arise on account of the use of Part to describe both type and instance, and if necessary any confusion can be resolved by referring to the type as a **Part-number**.

Parts come into existence as a result of **Demands** which form the bridge between the rest of the world and the Project. A Demand is a source of **Orders**, which it releases periodically into the Project. Demands are thus logically prior to Orders, and describe the ongoing processes which from time to time result in Orders being generated. An Order may be associated with one Part-number, and so a customer placing an order (with a small 'o') may generate several Orders. Another way in which an Order may come into existence is through a **Pull**, that is, when it is triggered by some change of state which occurs dynamically within the Alternative. Triggers of this sort are frequently generated by changes in levels of **Material**, a term which describes for the various inventory items which must be available to Machines in order to make the processing of Parts possible. These are often replenished automatically as a result of **Reorder Levels** being attained, or they can be refilled at pre-planned intervals by prescribing a **Delivery Schedule**. Material is also used to describe stocks of finished or partially finished Parts which find themselves in a state of temporary rest.

Each Order has a **Makespan** associated with it, that is a target completion time, after which it will be deemed to be **Late**. An Order may be made up of one or more Part instances. Once received into the Project, the Parts in an Order may be reconfigured into **Loads**, each of which consists of one or more Parts. A Load is the unit which is kept together throughout a sequence of Operations. The number of Parts in an Order is a matter for the *customer's* perception whereas the number of Parts in a Load is a matter for the *manufacturer's*. In the course of processing, Parts may be regrouped into **Batches**, which in turn may be regrouped into Loads. For example, Parts may be kept together in Loads according to Part type for much of their processing, but a wash process may be common to all Parts. For this operation Parts of different types could be regrouped into Batches. In general

7

they would retain their Part type identities for the purposes of unbatching in the course of further processing.

A simple diagram indicating the relationships between the principal modelling components described above is:-

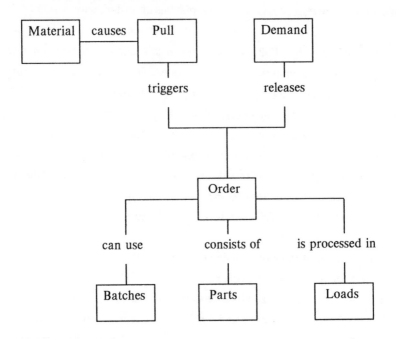

Figure 1. The role of Orders

Appendix A consists of a table showing the principal components of AIM.

1.4.1 Illustration

It may help to clarify the use of the terms in the above diagram by thinking of a hypothetical dairy producing cheeses. Its customers are the local cheese wholesalers who place purchase orders from time to time. These orders are the manifestation of a broader pattern or patterns of purchasing. Possibly each individual wholesaler has his own pattern, or it may be more convenient to think of a single universal pattern that transcends individual wholesalers and describes the whole ongoing time continuum within which the individual purchase orders occur at single instants. This is the concept underlying the ideas of Demand (continuum) and Order (instant). An Order is a single source of input, whereas a Demand is a source of sources.

Returning to the cheese factory there is a separate vat which contains rennet which is added to the main dairy product. From time to time the level in this vat falls to a minimum acceptable level and must be replenished. The rennet is a Material, and when it reaches its minimum level it causes a purchase action to be initiated by the dairy manager which eventually results in a supplier arriving to top up the vat. This is also an Order, but one triggered from within the dairy rather than without, and its source is a Pull. Alternatively the rennet supplier may operate on a regular Delivery Schedule.

The sizes of the orders placed by the wholesalers vary - some want dozens of different cheeses at a time, in which case a purchase order may generate multiple Orders. Others may want only one cheese or possibly even a fraction of a cheese. Within the cheese factory the foreman does not care at all about who is ultimately going to receive the cheeses; his care is that they should be processed in containers holding ten cheeses, all of which must be of the same type. The individual cheeses are the Parts and a group of ten cheeses is a Load. However, when it comes to the maturing process, there is no reason why the cellar should not contain a mixture of different types of cheese - that is, the cheeses are now processed in Batches - always assuming that they retain their type identity (Part-number) for the final stages of delivery to the customer.

1.5 Further Concepts and Terms

Whereas Parts and Material are the *passive* objects of manufacturing systems, the active agents are **Resources, Multi-Capacity Resources** and **Resource Groups**. A Resource Group is a group of Resources each of which fulfils the same role but whose members retain their individual identities, whereas the units of a Multi-Capacity Resource are indistinguishable. A series of identical robots identically programmed might exemplify the latter in the situation where a Part must be processed by just one member of the group. If on the other hand the robots were made by different manufacturers and had different manufacturing characteristics then a Resource Group would be appropriate.

Resource Groups are realised primarily in the form of **Machines** and **Operators**. Since Parts must move in order to be manufactured there is a general requirement for **Transport Systems**. There are three specific generic forms of transportation which occur sufficiently commonly in industry to merit explicit specification, namely **Transporters, AGVs** (Automated Guided Vehicles), and **Conveyors**. These can be thought of as special types of Resource which *move* rather than *transform* Parts and Material. The first two are organised into **Fleets**, and all three move along physical routes which are called **Transport Segments**. The term **Material Handling Device** is widely used in general industrial practice to describe these generic forms of transport, and this usage will be followed in this book, although it should be stressed that Material Handling Devices do *not* move Material.

10

Loads are rarely subjected to continuous activity, either manufacturing or transportation, during the time they are within a Project. Typically they have enforced rest periods during which they remain in a **Pool**, which can be broadly equated with the terms "buffer" and "WIP" (Work in Progress or Work in Process - there does not seem to be a general consensus about what the "P" stands for, nor does it seem to matter very much!). Usually a Load which allocates itself space in a Pool is also the Load which eventually frees that space, but this does not need to be so, and a **General Pool** is one for which this restriction is relaxed. To picture this distinction, a conventional car-park is a WIP - as each car is parked the available space is diminished and is not released again until the car is driven out. However, if the car park has reserved spaces which can become available by some action such as telephone calls from a distance by the drivers for whom they are reserved, this situation could be adequately described in terms of a General Pool.

Alternatives are **built** with two separate pictures in mind, the **Facility Window** and the **Process Plan**. The former is a stylised layout of the physical and material objects which would be found on the factory floor, that is, the machines, people, conveyors and so on. It is a *physical* plan in the sense of representing a real manufacturing layout in the style of a plan drawing. The latter is more akin to a flow chart in programming, and is discussed in detail in Chapter 2.

There is one component of the Facility window which is not physical in nature, namely a **Time-Persistent Value**. This is a value, often that of a **Variable**, which it is of interest to observe as an Alternative runs. It might be that of a count of objects passing through a certain point, or the utilization of a Resource. Time-Persistent Values can be thought of as conceptual

11

scoreboards and in some ways are real as the physical objects of whose performance they are a numerical abstraction. With real-time computers the devices on which they might be displayed could indeed possess a physical reality comparable with that of the other objects represented in the Facility Window.

By contrast the Process Plan is an entirely conceptual plan and consists of a series of **Jobsteps** which describe actions such as:

Operations (that is, activities which transform matter);
Setup (activities which make prepare Resources for Operation);
Transportation;
Route Selection;
Resizing of Loads.

The Jobsteps within a Process Plan describe the progress of Parts and Material whilst they remain within the Project, and also the behaviour of Resources, Pools and Material Handling Devices.

To complete the picture, Resources, Pools and Transport Segments are all subject to interruptions imposed by **Breakdowns**, **Maintenance** and **Shifts**. Shifts are time-patterns defining the availability and non-availability of collections of Resources, Pools and Material Handling Devices. The following diagram shows the inter-relations between the main terms which have been used so far.

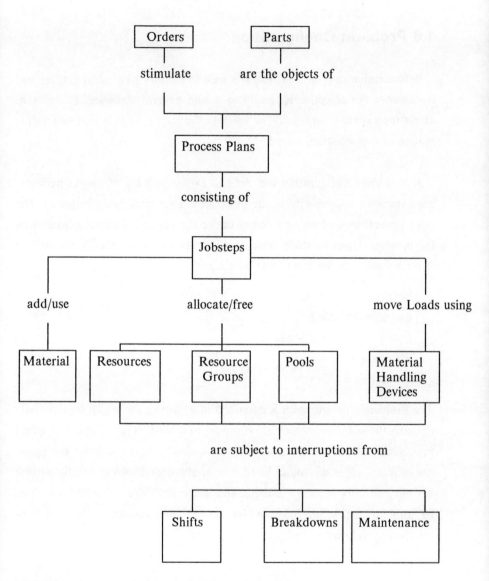

Figure 2. The role of Process Plans

13

1.6 Problem Configuration

Before embarking on a Project it is wise to identify its general nature, that is, whether the objective is simply to obtain general statistical information about the system under study, or whether the requirement is to obtain information of a more specific kind.

A Problem Configuration was defined earlier as a list of output performance statistics appropriate to the objectives of an underlying enquiry. The most general sort of enquiry comes under the heading **Summary Studies**. A list of other types of study within which the user can classify the sort of inquiries noted at the end of section 1.2 are:

Bottleneck Analysis;
Capacity Planning;
Material Handling Studies;
Throughput Studies;
Utilization Studies.

The Problem Configuration is described in a vocabulary which leans heavily towards those entities which are common to a wide range of manufacturing processes. Different collections of "interesting" output statistics pertain to the different types of studies listed above, although Problem Configurations are not inflexibly attached to these, and the term Problem Configuration can be used to denote any combination of available statistics. These may be grouped as follows:

14

Entity	Statistics
Order, Load, Part	Performance (see note below)
Resource	QL, WT, States (i.e. busy, idle, etc.)
Resource Group, Pool	QL, WT, UT
Material	QL, WT, Level
Conveyor Segment, AGV, Transporter Fleet	QL, WT, UT, States

QL = Queue Length ; WT = Waiting Time ; UT = Utilization

Performance = Observed Statistics of:-

Makespan (= Processing Time + Waiting Time)
Lateness (= actual Makespan - expected Makespan)
Processing Time
Waiting Time

where Observed Statistics is

(*average, standard deviation, minimum, maximum, count if appropriate*)

plus, for Parts only, average number in process.

In certain artificially simple circumstances quantities such as QL, WT and UT are connected by simple relationships, and the subject called Queueing Theory is concerned with what information can be gained by mathematical analysis as opposed to simulation in these cases. The rudiments of such analytic methods are discussed in Chapter 6.

The standard deviations of Makespan and Lateness are equal, since they differ only by the value of the expected Makespan. This is a constant for any given simulation run and can be defined as a parameter of a Pull or a Demand.

1.7 Demands

This is the primary mechanism for bringing new Parts into an Alternative and thus stimulating production. A Demand has five essential components, viz:

Part type;
Number of Parts to release at a time;
Number of Parts per load;
First release time;
Inter-arrival time (IAT).

First Release Time = 0 means that that the first instance of the demand occurs after the first drawing of an IAT.

The last item may be either a constant or a distribution indicating that the times between the arrival of individual Orders created by the Demand is drawn by random number generation from a statistical distribution. The most commonly used of these are (in alphabetical order) :

beta	lognormal	triangular
binomial	Normal	uniform
exponential	Poisson	Weibull
gamma		

and they are described in more detail in Chapter 3.

1.8 Processes

The term Process Plan has already been introduced in section 1.5. The underlying idea is that the physical entities of a manufacturing system become dynamic as a result of *processes* which are the direct and visible outcome of the work of engineers and designers. Processes therefore come about as a result of activities within human minds and brains. By contrast *mathematical* processes arise through the application of rules applied to *abstract* as opposed to *physical* entities. There is a sense in which mathematical processes exist even though nobody ever thinks about them or even discovers them! The most commonly referred to process of this sort in the world of simulation is the **Poisson Process**. A Poisson Process satisfies the conditions that objects arrive one at a time, independently of each other, and in such a way that the probability of an arrival in any time interval t is proportional to the size of t and does not depend on anything else in the system. Books such as Ref [10] contain proofs that these conditions lead with mathematical inevitability to an exponential distribution of inter-arrival times. Thus to say that inter-arrival times follow an exponential distribution is totally equivalent to saying that they are the outcome of a Poisson Process.

There is potential confusion concerning the relationship between Poisson Process and the exponential distribution on the one hand and the existence of a Poisson distribution on the other. The latter is a *discrete* distribution which determines the theoretical probabilities that the number of arrivals within a unit time interval is 0,1,2,... under the assumptions of a Poisson Process. The unit time interval is described by the mean number of objects arriving within it, and this is the sole parameter of the distribution.

1.9 Using AIM

This section is addressed to those with access to AIM, and contains instructions for building from scratch an Alternative (model), and an OS/2 DB2 Database to contain it. The version of AIM used in the following illustrations was 5.3.04. It is of course possible that variations may become necessary as the AIM product develops through further releases. Readers who are going to make serious use of AIM should of course have the relevant manuals - see Ref. [5].

AIM is a CUA (Common User Access) conforming product and it is likely that most users will be broadly familiar with the style of interface which this implies, particularly if they already have experience of windows using OS/2 or other systems. The following instructions will therefore assume a general familiarity with terms such as icons, minimize/maximize screen, dialog box, radio button, list box, etc. The first few steps are described in some detail. Thereafter it is assumed that dexterity in handling menus and windows increases rapidly, and thus more condensed instructions are appropriate. What follows is in the form of a tutorial with commentary. It consists of descriptions of the basic features of AIM interleaved with boxes which contain specific instructions for the construction of a trial

18

Database which exercises the main functions of the AIM program. Items which appear in bold print within the boxes indicate words or phrases which the user should look for in the appropriate window.

This book is not a comprehensive text on AIM. However, AIM contains ideas and constructs which support the notion of a common manufacturing vocabulary that will help simulation to become ever more prevalent and respected within manufacturing industry. AIM itself, like any other substantial software package, has many handy features which help its users become agile in communications through mouse and keyboard, for example the ability to cut and paste parts of one model into another. To discover the full range of accelerating aids such as these the user is referred to the AIM User Guide, Ref [5].

1.9.1 Installation Notes (version 5.3.04)

Prerequisites at time of writing:

An IBM compatible Processor capable of running OS/2. A Maths co-processor is recommended.

Colour monitor with appropriate OS/2 device driver. EGA minimum, VGA recommended.

Mouse with two buttons.

RAM: 10MB (minimum), 16 + MB recommended.

Installation space: 7MB for base product, plus a further 2.9MB to install all optional extras.

Installation is very straightforward. Put the first disk in the a: drive, type **a: install**, then answer questions and follow prompts. Three disks are required to install the base product. A fourth disk is also started by **a: install**. It contains the following further modules each of which can be separately chosen. None of these is necessary in order to follow this guide.

AIM Tutorial as described in AIM user manual	9Kb
Auxiliary data example	115Kb
Client/Server remote access utilities	170Kb
Example user code stubs	140Kb
Reference symbols catalogs	525Kb
Extended symbol sets (machines, conveyors, etc.)	920Kb

1.9.2 Project for the Trial Database

Parts arrive in a buffer from which they move into the first machine whose operation is known as making. They then move by conveyor to a series of identical test machines. Those which are accepted go direct to a final assembly machine, otherwise they go to a rework machine and then to final assembly. A scoreboard after final assembly records the total numbers which have passed through the system.

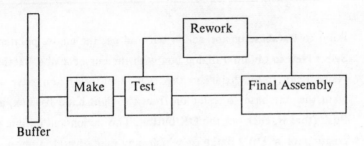

Buffer

Figure 3. Schematical representation of Trial Project

1.9.3 Loading the AIM program from the Desktop

Double click the left mouse button on the AIM program icon. After a while the AIM executive window appears which is labelled **FACTOR/AIM - Database:(Untitled)**

1.9.4 Loading/Creating a Database

To edit an existing Alternative or create a new one it is necessary to open the appropriate Database. The first choice is therefore whether to select **New..** to create a new Database, or **Open** to open an existing one.

Point to **Database** on the action bar and use the mouse pointer to Select **New** to obtain a dialog box with the cursor in the **Database Name** field. Enter "trial" in this field, then select the drive you want the database to reside on from the right hand list box, and Push (that is, click on) the **OK** button. This action will initiate the creation of a DB/2 database which may take several minutes. At the end of this time a new window called **Alternatives** appears within the executive window.

An alternative action from this window is to highlight the name of one of the listed Databases, click the left mouse button on it, then either click on OK or hit the return key.

1.9.5 Logging on to the Database

Depending on the underlying database management system you may be required to log on to the Database. Following this, the Database is loaded which again may take an appreciable amount of time.

1.9.6 Selecting Alternatives

Select **Simulate** from the **FACTOR/AIM** action bar, then select **Build/Run** which causes a **Build New Alternative** window to appear.

Enter "Base Case" in the **Description** field and Push **Build**.

In order to make the descriptions of user actions less verbose a convention will be adopted whereby a line such as the last one in the box above is abbreviated to the line

Description : Base Case ; Push **Build**

in which the semi-colon is used to separate two distinct actions.

One or more alternatives may be selected from the Alternatives window. A single Alternative is selected by double clicking on it. If more than one Alternative is listed multiple Alternatives are selected by holding down the Control key while doing multiple clicks - this feature will become important when the results of two or more Alternatives are available for comparison.

1.9.7 Setting Build Options

The options available fall under the headings:

Background (that is, colour and bitmap);

Preferences (that is, whether or not Facility window has a grid);

Input (that is, Base or Extended - see below).

Once an Alternative is loaded it is a good idea to use the **Build..** menu to change the input edit option to **Extended**. The Base level of editing is restrictive, and many of the examples in this text use facilities which are only available in the Extended mode.

From the **Simulator** window action bar select **Options**, then **Build** from the pull-down menu. Select the **Extended** radio button, then Push **OK**.

1.9.8 The Simulator Windows

There are three main windows called Facility, Simulation Status and Simulation Trace. The last two are operative only when simulations are run. It is possible to maximise the Executive window which in turn maximises the Facility Window thereby providing a bigger canvas for adding graphical elements to the Alternative.

1.9.9 Adding Graphical Elements

Graphical components are added using the Add Components window which is opened by selecting the **Edit** and **Add..** options. This window is moveable but not rescalable. To move it, click on the title bar of the window which contains the words **Add Components** and use the mouse to drag it to a position on the screen where it does not interfere with the part of the Facility Window you are currently working on. Add a machine by clicking on the machine icon. The mouse pointer then turns into this icon and can be moved around the window. Once its desired location is reached double clicking the left mouse button places the machine there. This approach works for the following graphical components: Machines, Operators, Fixtures, General Resource (in all cases both single and multiple), also for Material and Time-Persistent Values.

Once an addition has been made, the pointer is dragged over the arrow icon within the Add Components window and clicked, which returns the pointer to normal. It is then possible to move an icon on the Facility Window by dragging it with the left button depressed, and releasing it in the desired location.

Create objects within the Facility window which realise the items shown in Fig. 3. From the Alternative window action bar select **Edit**, then **Add..** from the pull-down menu.

By placing the cursor within the title bar, drag the **Add Components** window to the top right hand corner of the screen.

Add three machines corresponding to Make, Rework and Final Assembly in Fig. 3. and place them in their correct relative positions. At this point the Facility Window will contain just three squares as shown in the following diagram.

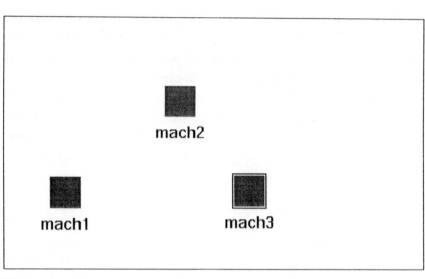

Figure 4. Add Components Window

Add the Multi-capacity machine (Test) and place it between mach1 and mach3. Double-click on the icon. This causes an editor screen to be displayed, in which change the value in the **Capacity** field to 2. Then push **OK** to return to the Facility window.

In all cases accept the default names for the objects created, that is, mach1, mmac1, etc.

Transport Segments and WIP Pools work in a slightly different way. An icon is selected as above, dragged to the start position and clicked. Then the

26

pointer is moved to the end position and a double click establishes the other end. Finally the cursor is dragged over the arrow box and the mouse button clicked to disengage.

> Using Fig. 3 as a guide, incorporate the buffer by adding a WIP Pool as described in the immediately preceding paragraph.

1.9.10 Creating Transport Segments

These are like WIPs but additionally multiple segments containing turns and corners can be defined within a single screen placement operation.

> Add a single conveyor segment connecting mach1 and mmac1. Observe that two **Conveyor Control Points (ccp1** and **ccp2)** are automatically created at the ends of the segment.

1.9.11 Add Time-Persistent Values

In order to display information which is of value to the user, Time-Persistent Values must be associated with the values of **Variables** which are updated by assignment at run time. It is thus natural to construct Time-

Persistent Values and Variables in pairs. Variables by their nature are **non-graphical** objects.

Add a **Time-Persistent Value** using the **Add Components** window and place it somewhere near mach3. Double-click on the icon in the Facility window to reveal the Time-Persistent Value Editor. In the **Description** field enter the text "tput=" .
  Push **OK**.

Use the list box at the bottom of the **Add Components** window to select **Variable**. You may either use the mouse along with the slider on the right to make the selection or click on the pull-down icon and type the character 'v'.
  Push **Create**.

From the **Type** section select the radio button **Integer**. Push **OK**.

Edit the Time-Persistent Value by Double-clicking on the Facility window. Select **Expr** : var1 Push **OK**.

The result should be a Facility Window like that below which bears a reasonable likeness to Fig. 3. on which it was based.

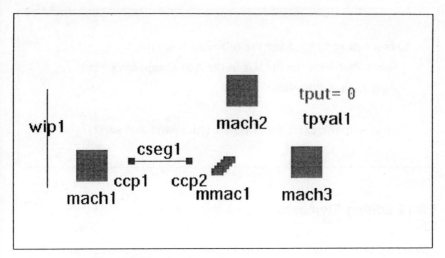

Figure 5. Facility Window for Base Case

As a refinement you could double-click on each of the machines and the wip in turn and edit their **Name** fields to give them names resembling those in Fig. 3. You are allowed up to eight characters in each name.

1.9.12 Creating non-graphical Components

The procedure described in the box above applies more generally to the creation of all non-graphical components, that is, from the Add Components window, select from the list of available entities and click on **Create**. This results in the display of an appropriate editor (one editor per element type). There is thus a plethora of different types of editor.

Create two Parts by doing the following twice over:
Select **Part** from the list-box in the **Add Components** window.
Push **Create**. Push **OK**.

You have now created the two Part types **part1** and **part2**.

1.9.13 Editing Elements

Once an element has been created subsequent changes must be made by editing. Select **Edit** from the executive window action bar, and then from the **Edit** pull-down window select **Find/Select**. For graphical elements it is also possible to proceed directly to edit by double-clicking on items in the Facility Window.

1.9.14 Deleting (Clearing) Elements

For graphical items double-click on an item in the Facility window, then either press the **Delete** key on the keyboard, or alternatively, select the **Edit** pull-down window, and then select **Clear**. For non-graphical items, obtain the appropriate editor by selecting **Edit**, then **Find/Select**. Then click again on **Edit** in the action bar and from the **Edit** pull-down window select **Clear**.

1.9.15 Demands

Demands and Pulls, which are discussed in detail in section 5.1, bring new

Parts into a system in an ongoing fashion and are thus prime movers without which nothing would ever happen in execution. They are not themselves part of the Process Plan but are an essential precursor of it.

Select **Demand** from the list-box in the **Add Components** window and create two Demands. From the editor screen fill in details as follows:

dord1:
  Number of Parts to Release : 4
  Number of Parts per Load : 1
  First Release Time : 0
  Inter-arrival Time : exponential(1)
  Part : Part1 (chosen from list-box)
  Push **OK**

dord2:
  Number of Parts to Release : 6
  Number of Parts per Load : 1
  First Release Time : 0
  Inter-arrival Time : exponential(1.25)
  Part : Part2
  Push **OK**

It is also possible to impose an overall maximum on the number of Orders released by a Demand.

31

1.9.16 Saving Work

The end of a chapter is a natural break-point at which to pause and save work with a view to resuming at a later date.

From the Alternatives Window select **File** ; **Save**

(Alternatively from the Alternatives Window select **File** ; **Exit**.)

A Save window appears from which check **Unload Alternative**, then Push **Save**.

2. PROCESS PLANS

2.1 The need for a Process Plan

The objects discussed in Chapter 1 are either *physical* (Machines, Conveyors, etc.) or *generative* (Demands and Pulls). The separation of Product and Process is often made when analysis of manufacturing operations is undertaken. Underlying this is a distinction between physical things (Product) and abstract things (Process). A Process is like a computer program in having no physical existence in its own right other than that of the medium needed to record it. Process is a generic word describing the manner in which physical objects interact to transform matter. By analogy, in simulation, once the physical objects have been laid out and the pattern of demand specified which is the stimulus for the transformation of matter, nothing more can happen until "Process" is applied. Clearly no manufacturing process can be comprehended, let alone analysed, in one fell swoop. It must be broken down into manageable steps in the same fashion in which a computer program breaks down an algorithm into single statements in a programming language.

The analogue in simulation of the programming language statement is the **Jobstep,** and just as there is a wide variety of statement types in any of the high level programming languages, so there is a variety of types of Jobstep which can make up a Process Plan. Almost all of them include the activity of allocating and freeing Resources. Those which do not,describe *decision* activities; some Jobsteps such as Inspect involve both of these types of activity.

2.2 Project for the Trial Database

The informal description of the project is repeated:

Parts arrive in a buffer from which they move into the first machine whose operation is known as making. They then move by conveyor to a series of identical test machines. Those which are accepted go direct to a final assembly machine, otherwise they go to a rework machine and then to final assembly. A scoreboard after final assembly records the total numbers which have passed through the system.

This can be translated into a sequence of Jobsteps, each of which is roughly comparable to a statement in a high-level programming language.

It is at the level of the Jobstep that numerical values are applied. All programming languages contain the concept of **Variables,** and in an Alternative
it is at the Jobstep level that these would be used and **assigned.** AIM Variables are by definition global in scope; a value which is assigned to a particular Load is called an **Attribute.** Since Variables have different values in the
course of simulation runs it is possible to associate **Observed Statistics,** that
is, mean, standard deviation, maximum, minimum and, where appropriate,
count, with each one.

Often it is useful to store all or most of the values of Variables and Attributes used within an Alternative in the same place for ease of editing. Such
a repository of numerical values is called a **Lookup Table.** The essential elements of a Process Plan can be represented as:

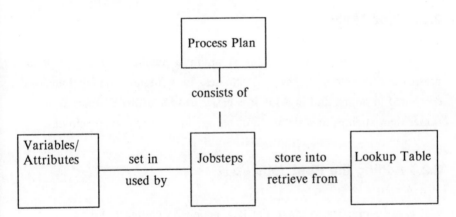

Figure 6. Elements of a Process Plan

2.2.1 Trial Project Jobstep List

Initial analysis of the problem leads to the construction of a Process Plan with the following Jobsteps:

Jobstep 1 : Move parts into Buffer
Jobstep 2 : Make
Jobstep 3 : Convey from make to test
Jobstep 4 : Test
Jobstep 5 : Decision following test
Jobstep 6 : Rework
Jobstep 7 : Final Assembly
Jobstep 8 : Count throughput

2.2.2 Time Units

In some simulation systems time is simply a number and can be interpreted in whatever units the user chooses. In AIM, however, all times are computed in hours, and in AIM it is easier to standardise all times in hours rather than attempt to scale by "pretending" that an hour is something else.

2.2.3 Adding the First Jobstep

It is good practice to make the first action on editing a Jobstep that of filling in the **Description** field. This is an informal description using words such as those in the Project trial Jobstep list in section 2.2.1, and forms the primary part of the user's own documentation.

The most important consideration in editing Jobsteps is that of **allocating** and **freeing** Resources which have already been defined in building the **Facility**. Every Jobstep begins with an allocation stage during which resources are allocated and freed. The words **before** and **after** in the options **Free Before** and **Free After** mean before and after this allocation phase as opposed to the Operation Time involved in the main body of the Jobstep.

The stages of Jobstep processing are:

1. **Free Before** (i.e. before any Resource allocation);
2. **Perform job-specific tasks** (e.g. aggregation - see section 7.2 and allocation of Tags for Kanbans - see section 5.2);
3. **Allocate Resources** (with possible delay if Resources not immediately available);
4. **Free After** (i.e. after any Resource allocation but *before* the work of Jobstep);
5. **Passage of Time** required for Operation, Move or Setup;
6. **Free End**;
7. **Perform job-specific tasks**.

Create a Process plan by selecting **Process Plan** from the list-box of the **Add Components** window. You can either use the slider and mouse, or click on the pull-down icon and type 'p' three times. Then push **Create**.

Push the **Insert** button.

From the **Select Jobstep Type** window select **Setup/Operation** which is the default. Push **OK**. The Jobstep editor for js1 should be displayed.

Complete the **Desc** field using the appropriate phrase from the Project trial Jobstep list above, which is "Pull parts into buffer."

Jobstep Specification may involve a fairly large amount of input data, but many fields are given default values which are often adequate, at least for the first few runs. The various Jobstep editor windows are divided into sections, and often continue through a **More..** prompt into a further window. The main sections of such windows are associated with labelled group boxes.

There are three group boxes in the js1 editor window :

Resource/Pool/Group

Operation Time

Set-up

Make entries in the first of these only. Use the three small fields on the left which are called **Name**, **Units**, and **Action**. (The first and last of these are list-boxes.) Use these lists to create an entry in the right hand list-box as follows :

wip1   1   Allocate

This is an instruction to allocate one unit of wip1, that is one position in the buffer.

Push **OK**

38

2.2.4 Adding an Operation Jobstep

The second Jobstep is **Insert**ed by the following:

js2: From the **Process Plan Editor** window push **Insert**, then push **OK**.
Complete the **Description Field** from the Project trial Jobstep list, namely "Make."
In the **Resource/Pool/Group** section use the three small fields to create the entry
   wip1 1 Free After
Push **New** to the right of the main entry box, and create another entry:
   mach1 1 Allocate
Operation Time : normal(.1,.02)

Push **OK**

The Operation Time given in the box above invokes the Normal distribution - an account of distributions which may be selected is given in Chapter 3. Other possible ways of expressing Operation Times are as Loads per hour, Parts per hour, or Part Cycle Time.

2.2.5 Allocation

A Jobstep may contain a list of Resources to be allocated. Usually, as in Base Case, this means that the Jobstep requires *all* of the Resources in the list to be allocated, but it is also possible that just one is sufficient. It is also possible to specify that Materials are needed before the Operation part of the Jobstep can proceed.

If the requirement is **All** various other questions must be addressed, viz:

must the resources be allocated in the order (that is, sequence) given in the list or is any order allowable?

must they be simultaneously available for allocation as a group or can they be allocated as they become free?

is **Order Integrity** required, that is, are all Loads of one Order to be completed before the next is started?

is **Part Integrity** required, that is, are all Parts of one Load to be completed before the next is started?

2.2.6 Adding Subsequent Jobsteps

The third Jobstep relates to the movement of Parts from Make (mach1) to Test (mmac1). Two types of Move can be distinguished, the first involving a further agent such as a Material Handling Device, for example a Conveyor or AGV (Automated Guided Vehicle). This type of move is described by a **Move-Between** Jobstep.

The second type of Move is when no further agent is involved but time must elapse between the part being at two different physical locations, for example a part is lifted by a human carrier from one point to another, but the carrier is not explicitly modelled in the Alternative. This is sort of movement is like an Operation without Resource allocation. It is described in a **Move** Jobstep.

Since the movement from Make to Test is carried out by a Conveyor it should be modelled by a **Move-Between** Jobstep rather than a **Move**.

The instructions for the third and remaining Jobsteps are given in a slightly abbreviated form.

---

js3 : Push **Insert** and a **Select Jobstep Type** window appears.
Push **OK**
In the window use the **Type** list-box to select **Move-Between**.

In the subsequent **Extended Move-Between Editor** window do:
In the **Type** section push the **Conveyor** radio button.
Conveyor System = csys1 ; Origin = ccp1 ; Destination = ccp2
Free at Pickup = mach1 ; Allocate at Drop-off = mmac1
Push **OK** ; Push **Insert**

js4 : Select **Set-up/Operation** ; Push **OK**
In the **Resource/Pool/Group** section enter
mmac1 1 Free End
Operation Time : uniform(.2,.3) ; Push **OK** ; Push **Insert**

41

js5 : Select **Probabilistic Select** from the Process Plan editor
list-box ; Push **OK**
In the Probabilistic Select box do:
Jobstep : js1   ;   **Probability** : 1   ; Push **OK** ; Push **Insert**

js6 : Select **Set-up/Operation** ; Push **OK**
In the **Resource/Pool/Group** section enter
    mach2 1 Alloc/Free
Operation Time : normal(.5,.2)   ; Push **OK** ; Push **Insert**

js7 : Select **Set-up/Operation** ; Push **OK**
Find the **Resource/Pool/Group** section and enter
    mach3 1 Alloc/Free
Operation Time : .1 ; Push **OK** ; Push **Insert**

js8 : In the **Select Jobstep Type** window
use the **Type** list-box to select **Assign**.
Push **OK**

Within the **Extended Assign Jobstep Editor** find the **Assign** section
and within it:
    set the **Type** radio button to **Variable**
    within the **Name** list-box select **var1**
    in the **Expr** field type **var1 + 1**   ; Push **OK**

It is now necessary to revisit the Probabilistic Select Jobstep, since the
Jobsteps to which it must refer were not created at the time of its own creation.

Select js5 by double-clicking within the Process Plan Editor.

Edit the existing entry to make it:
Jobstep : js6; **Probability** : .1

Push **New** to create a new entry:
Jobstep : js7; **Probability** : .9

Push **OK**

At this point the main field in the Process Plan Editor can be used to click on any of the individual Jobstep lines and do any further editing or corrections which may be necessary at this point. You should be satisfied that you have entered a correct Process Plan before you proceed to the next box.

From the Process Plan Editor push **OK**.

Close the **Add Components Window** by clicking top left and selecting **Close**.

2.2.7 Checking the Model

Once the first draft of an Alternative is complete it can be tested for consistency by selecting **Utilities** in the executive window, and then **Check Model**.

Select **Utilities** from the Alternative action bar.

Select **Check Model**.

This should show at least two errors corresponding to each of the two Demands. (If there are more than two errors you must use a combination of intelligence, ingenuity and the technique below in order to eliminate them!)

Double-click on each of these errors in turn to enter the Demand Detail Editor. In each case do:

  Select **Process Plan** : pplan1

  Expected Makespan : 0.5

  Push **OK**

The last entry says that any Order which remains in the Project for more than 30 minutes is deemed late.

Push **Done** ; Rerun **Utilities** ; **Check Model** ; Push **OK**

After the check is run either a message appears to indicate that the model is good, or a window appears listing all the problems found. Double-clicking on any of these causes the appropriate editor to appear together with an indication of what information is missing or needs to be added.

At the Check Model stage the built-in help screens can be useful. Often a whole list of problems arise from only one fault. For example if a Control Point is deleted, all items referring to that Control Point will produce an error line. It may be necessary to use the horizontal scroll bar to read the error message in full.

3. DISTRIBUTIONS

3.1 Choosing Distributions

The need to supply parameters for Alternatives frequently leads to a period of observation and investigation in which a distribution is fitted empirically. Whether a discrete or an empirical distribution should be fitted is inherent in the nature of the type of simulated data which is to be generated. In the case of continuous data the simplest distribution is uniform. When this is clearly unacceptable the next simplest choices are Normal if the data is peaked and looks roughly symmetrical, and exponential if it is heavily skewed and tails away from a peak at the left hand end.

A table of the most frequently used distributions along with their parameters is given below:

Discrete: parameters

uniform	min,max
triangular	min,mode,max
binomial	N,p
Poisson	mean

Continuous:

Normal	mean (μ), standard deviation (σ)
lognormal	μ, σ (see section 3.3)
exponential	α (scale) $= \dfrac{1}{\mu}$
Weibull	α (scale), β (shape) (n.b. $\alpha \neq \dfrac{1}{\mu}$)
gamma	α (scale), β (shape) ($\alpha \neq \dfrac{1}{\mu}$)
beta	p,q (see equation below)

Sometimes it is the case that previously collected data is available to form an *empirical* probability distribution. In this case aggregated data rather than parameters can be used for random number generation (see section 3.5), and this aggregated data is most conveniently stored in a Lookup Table. The data concerned may be either discrete or continuous. For example, a count of faults is necessarily discrete, inter-arrival times would probably be continuous.

3.2 Exponential, Weibull and Gamma distributions

The equation of the unscaled exponential distribution (or to be somewhat more exact, the equation of its probability density function) is $y = e^{-x}$, an equation which can hardly be bettered for simplicity!

Probability density, which is usually plotted on the vertical axis, is for continuous functions what probability is for discrete functions, that is it is probability per unit of x where x is whatever is measured along the horizontal axis. Thus, unlike probability which is dimensionless, probability density has the inverse dimension of the x axis unit of measure.

The Weibull and gamma distributions are different extensions of the exponential distribution, each of which gives rise to a *family* of distributions. The parametric equation of the unscaled Weibull family is

$$y = \beta x^{\beta - 1} e^{-x^{\beta}}$$

and that of the gamma family is

48

$$y = \frac{1}{\Gamma(\beta)} x^{\beta-1} e^{-x}$$

again equations of remarkable simplicity, particularly in view of the fact that each of these families contains distributions which cover an enormously wide range of forms.

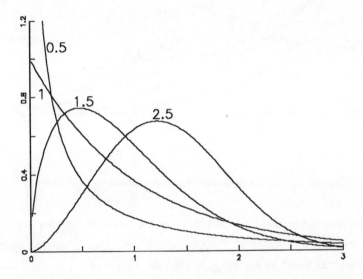

Figure 7. Weibull curves for different shape parameters

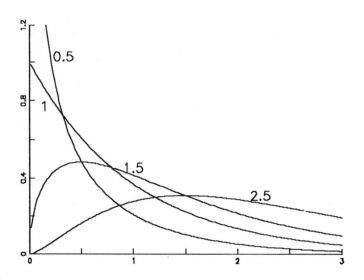

Figure 8. Gamma curves for different shape parameters

The equation of the gamma family involves the so-called Gamma function denoted by $\Gamma(x)$, a continuous function which coincides at integer points with the factorial numbers according to the rule $\Gamma(n) = (n-1)!$.

Many packages include the so-called Erlang family in their list of distributions. These (named after a Danish telephone engineer) are simply the special cases of the gamma family for which β is an integer.

The equation of the unscaled beta curves is

50

$$y = \frac{1}{B(p,q)} x^{p-1} (1-x)^{q-1}$$

where $B(p,q)$ is the so-called Beta function defined as

$$B(p,q) = \int_0^1 x^{p-1} (1-x)^{q-1} dx$$

To scale the exponential, Weibull and gamma distributions replace x by $(\frac{x}{\alpha})$ and multiply the right hand side by $\frac{1}{\alpha}$. The exponential distribution is the special case $\beta = 1$ of both the Weibull and gamma distributions.

The following table gives the means and modes of these distributions, and further emphasises the way in which the second two are straightforward developments of the first.

	mean	mode
exponential	α	none
Weibull	$\alpha\Gamma(1 + \frac{1}{\beta})$	$\alpha(1 - \frac{1}{\beta})^{\frac{1}{\beta}}$
gamma	$\alpha\beta$	$\alpha(\beta - 1)$
beta	$\dfrac{p}{p+q}$	$\dfrac{p-1}{p+q-2}.$

Note: Weibull and gamma distributions only possess modes if $\beta > 1$.

The beta distribution is particularly useful when data to be fitted is clearly finite in both tails. The uniform distribution is a special case of the beta distribution with both parameters equal to one. A triangular distribution with a peak at one or other extreme is also a particular case of the beta distribution, as is the arcsin function.

Information about means and modes can provide useful rules of thumb in making first judgements about fitting curves to data. The following is a technique for fitting a beta distribution when the data is both finite and single-peaked.

Scale the data so that its maximum value maps to 1 and its minimum value to 0, and compute the scaled values of the mean and mode. Substituting these in the formulae for mean and mode given above leads to two simultaneous linear equations which can be solved for p and q.

For example, suppose some data has been obtained which is skewed and single-peaked, and has minimum and maximum values 3 and 7 respectively. Suppose further that the mean is 5.5 and the mode (peak) is at 6.

Scaling the data to the range (0,1) adjusts the mean and mode to be 0.625 and 0.75 respectively, so write

$$\frac{p}{p+q} = \frac{5}{8}; \quad \frac{p-1}{p+q-2} = \frac{3}{4}.$$

The solution of these equations is readily found to be $p = 2.5, q = 1.5$, so select these values as the parameters to fit a beta distribution.

3.3 The Normal and lognormal distributions

The equation of the Normal distribution with parameters μ (mean) and σ (standard deviation) is

$$y = \frac{1}{\sigma\sqrt{2\pi}}\, e^{-\frac{(x-\mu)^2}{2\sigma^2}}$$

Since the Normal extends indefinitely at both ends it is often wise to fit a *truncated* Normal distribution to observed data.

The apparent complexity of the equation for the Normal distribution in the form given above can be off-putting, and it is worth while quoting its *differential* equation for the case $\sigma = 1, \mu = 0$ which is:

$$y' = -xy$$

in order to see that it has all the simplicity ascribed above to the exponential distribution.

The equation of the lognormal distribution with parameters μ (mean) and σ (standard deviation) is

$$y = \frac{1}{x\sigma\sqrt{2\pi}}\, e^{-\frac{(\ln x - \mu)^2}{2\sigma^2}}$$

In an abbreviated phrase which sacrifices strict accuracy for the sake of ease of recall, lognormal means that if x is lognormal then log x is Normal. The word "Normal" in the context of the Normal distribution is spelt with a capital to distinguish it from the "normal" use of the word. No such distinction need be made for the word "lognormal."

53

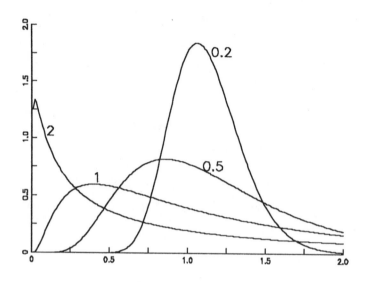

Figure 9. Lognormal curves for different values of σ

The interpretation of the parameters of the lognormal distribution requires a little care. As the above diagram shows σ determines the shape of the distribution. It is easy on first introduction not to spot the x on the denominator. Its presence means that the value of μ, which in the case of the Normal distribution measures solely location, now measures a mixture of location and scale. Also y is not defined for $x = 0$, but it is always the case that $y \rightarrow 0$ as $x \rightarrow 0$ however great the value of σ, a point which is illustrated by the small peak at the left of the $\sigma = 2$ curve in the diagram, which illustrates how steep the initial ascent is for higher values of σ. The parameters

54

μ and σ estimate the mean and standard deviation of the "logged" values of x. However some statistics packages return the **anti-log** of μ so that the estimate returned is comparable in magnitude with the original values of x.

Algorithms exist for making fits to such distributions, that is, finding the "best" set of parameters according to criteria established by statistical theory. Such algorithms are readily found in statistical packages, and so when simulation involves making estimates of numerical parameters from observations made on the ground, it is a prerequisite that an appropriate statistics package is also to hand. Such a package will deliver not only parameter estimates, but also confidence intervals for them. Any one set of observations of repeated values such as times taken by an operator to perform a task, or historic records of breakdowns on a set of equipment is no more than a single sample, and it is usually the case that the best thing to do is to select an integral or well-rounded parameter value which lies within the confidence interval.

For skew distributions the first consideration is to identify the best-fitting structure, then find the best-fitting parameters within that structure. The set of lognormal, Weibull and gamma curves provides different structural forms, thereby increasing the possibility of obtaining a good fit.

3.4 Estimating distributions and parameters

This section might be subtitled the story of Karen and Julie-Ann, two young ladies who agreed that their operation times in the rework area of a factory be measured for the purposes of input to a simulation model. The first few readings are as follows:

Karen	Julie-Ann
1.25	3.00
0.83	1.33
1.00	1.03
1.16	1.24
0.75	1.00
1.05	1.30
0.83	0.86
1.70	1.26
1.00	1.10
0.85	2.00
0.83	0.86
1.40	1.63
1.15	1.30
0.67	1.56
0.86	0.95
...	...

A first step prior to curve-fitting is to look at the data in Exploratory Data Analysis fashion, with a particular eye to judging whether any of the data items should be excluded either on the grounds of their being outliers, or because they give evidence that there is a second and separate process at work of which explicit account should be taken. Histograms of the four sets of data along with Weibull fits are shown below:

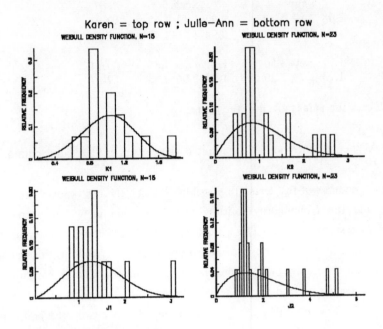

Figure 10. Operation times with Weibull fits

The four distributions are all peaked with a strong left skew but with a small sprinkling of low values which would allow any of lognormal, Weibull or gamma to be considered. The distributions were subjected to analysis in the curve-fitting component of a well-tried statistics package. Each girl was timed over two sequences of operations and the resulting Weibull fits were as follows:

57

	Shape	Scale	χ^2
Karen	2.48 3.89 5.29	0.97 1.12 1.28	3.97
	1.34 1.90 2.47	0.94 1.21 1.49	2.78
Julie-Ann	1.71 2.60 3.50	1.21 1.53 1.85	3.14
	1.23 1.73 2.23	1.54 2.06 2.58	20.7

In each case the values given in the order:

lower 95% confidence bound,   estimated value,   upper 95% conf. bound.

Various "goodness-of-fit" tests are possible of which the best-known are the χ^2 test and the Kolmogorov-Smirnov test, discussion of which properly belongs to a statistics text. The χ^2 values given above should be compared to the column in χ^2 tables appropriate to one degree of freedom, this being one less than the number of parameters being estimated. The χ^2 test should be used in a more relaxed way than is customary in other sorts of statistics applications. For one thing the χ^2 value obtained depends on the boundaries chosen for allocation of values to cells, and so there is no single absolute value of χ^2 which measures the goodness of fit of a distribution. Simulation in general is not like those areas of science in which high precision of measurement is an over-riding goal. In simulation it is only large effects which matter, which means that in matters such as judging goodness of fit, visual examination is often little inferior to scrutiny of χ^2 values. This is not say that one would fly in the face of an outrageously large χ^2 value, but it does mean that one should not be pedantic in terms of interpretation of 95% or any other cut-off point.

The present data suggests that Julie-Ann's times have a more "exponential" sort of shape than Karen's, and also that her second sample is more strongly skewed to the left than the other samples. It has also to be borne in mind that Karen and Julie-Ann are themselves two samples from a wider population of all the rework girls, and if, as is often the case, a single parameter set is required which seems not unreasonable for covering all cases, then Weibull (2,1.25) might not be too far off, even in spite of the high χ^2 value for Julie-Ann's second run, which clearly arises from a degree of compression in the middle of the data and a few aberrantly high values.

Values for lognormal fits were

		μ			σ	
Karen	-0.148	-0.009	0.130	0.184	0.243	0.396
	-0.290	-0.070	0.150	0.393	0.497	0.720
Julie-Ann	0.065	0.250	0.435	0.244	0.323	0.527
	0.226	0.448	0.669	0.396	0.500	0.725

To illustrate the remarks in section 3.3 about exactly what the lognormal parameters estimate, the means and standard deviations of the *actual* times recorded are:

Karen	(1.022, 0.274)	(1.066, 0.620)
Julie-Ann	(1.361, 0.548)	(1.813, 1.194)

and the means and standard deviations of the *logarithms* of the times recorded are:

Karen	(-0.0088, 0.251)	(-0.070, 0.508)
Julie-Ann	(0.250, 0.334)	(0.448, 0.512)

3.5 Random Number Generation

All simulation packages provide the possibility of using several different random number streams, thereby enabling the same Alternative to be run with what are effectively different scenarios, and also providing the opportunity of running different scenarios with the same fundamental source of input.

Generation of a random number stream from an initial number known as a **seed** consists of performing arithmetic within a computer according to one of a range of algorithms which vary widely in complexity. The quality of random number generators is a much researched area in Numerical Analysis. An account of this and of the details of many algorithms for generating random numbers can be found in Ref. [6].

Since a computer can start from the same seed and use the same algorithm, random numbers streams are repeatable, and thus not truly random. For this reason they are often talked about as generating **pseudo-random numbers**.

Many simulation users will never need to know any more about random numbers than this, and in particular do not require to know about the sort of mathematical processes by which random numbers from the various distributions are generated. However, the primary model for drawing random numbers from an arbitrary distribution is so simple that it should be appreciated by all simulation analysts.

The technique consists of plotting the **distribution function,** that is the *cumulative* probability density function of the distribution, as y, versus the range of all possible values of the variate as x.

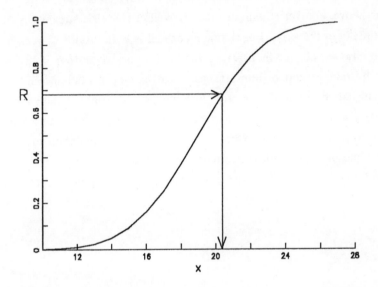

Figure 11. Random number generation

The y values thus range from 0 to 1, and the generation technique consists of drawing a random value R from the uniform distribution (0,1), projecting horizontally until the curve is reached, and then projecting vertically down-wards to the x-axis where the random value from the distribution is read off.

3.5.1 Antithetic random number streams

Continuing the illustration in the above section, if a stream of random uniform numbers R is used to generate a stream of an equal number of say random Weibull values, then replacement of every value R by (1-R) produces another random uniform stream from (0,1) and thus, using the technique above, another stream of random Weibulls. Such a stream is said to be **antithetic** to the first. For simple models, it is both intuitively reasonable, and also borne out in practice, that the values of output statistics obtained by taking random numbers generated half by a random number stream and half by its antithetic are more narrowly distributed than those generated by a single stream. A narrow distribution means that outputs can be estimated more accurately, which is beneficial to the simulation overall. The technique of using antithetic streams is a special instance of variance reduction (see section 8.3). Antithetic variables are not available in AIM.

4. RUNNING SIMULATIONS

4.1 The run-time environment

Run-time can begin when the Facility Window (that is, the physical lay-out) and the Process Plan have been completed to a sufficient extent that the Alternative can be started up in mimic mode. Typically there is then a period when the user performs repeated runs with intervening switches to modelling in order to be satisfied that the Alternatives within the Project accurately satisfy the the model specifications as laid out in the the initial requirements laid down either by himself, or more probably by a client. This is a process of iterative refinement with the objective of model **Verification**. When this stage is complete the next stage is called **Validation**, that is, confirmation that the model is indeed very like the reality it sets out to copy. At this stage the factory managers or the foremen or the group leaders may be brought to the computer terminal, hopefully to confirm the likeness and thereby admire the modeller's skill, or at the least to learn something from the messages which the model conveys. It is hard to generalise about the expectations and reactions of clients at this stage. Certainly presentation and "cosmetics" such as animation and dynamic graphs are powerful forces and should not be skimped on (see also section 8.1).

The environment in which simulations are run can be tailored to give different styles of animation. Extremes would be running the animation at the same speed as reality on the one hand, or at the other updating the screen only at the end of the simulation. Between these extremes indefinite graduations are possible. On a run of a simulated week, for example, screen updates every simulated ten minutes may be necessary in order to give an impression of smooth continuous movement. On the other hand there are circumstances in which it is sufficient to update only at the end of shift or end of day, and display whatever results are desired.

Sometimes it may be best to switch off animation altogether, thereby enabling the computer to progress at a much faster speed, perhaps through many weeks of simulated time. Running without animation is called **batching**, which should not be confused with batching of Loads as described in section 7.2.5. It is not unusual for the length of such a run to exceed in simulated time the whole lifetime of the real process so that long run averages of quantities such as utilization, queue lengths and service times are well estimated. Later chapters deal with these considerations in greater detail - it is sufficient at this point to indicate the many choices and options which are available to the user at run-time.

4.1.1 Simulation options in AIM

The following box continues the example from the end of Chapter 2. Start and finish times are entered for the simulation run, and optionally a time can be declared at which statistics should be cleared in cases where it is decided to run an Alternative with a **Warm-up Period** in order that the statistics which are gathered should pertain to conditions of steady state.

Select **Options**, then **Simulate..** from the Alternative action bar
Start Time : 0 ; **Finish Time** : 3

Push **OK**

When a simulation is run, a Simulation Trace window can be displayed which can be useful both in initial debugging, and also in refining an Alternative or studying some parts of one in small detail.

4.1.2 Starting and Stopping a Simulation Run

Once the simulation period and animation options have been set, a simulation is run by selecting the **Simulate** option from the title bar in the Facility Window. For the first run of an Alternative select **Go**.

The Alternative is checked and the animation begins. At any time in its course the **Stop** option from the **Simulate** menu can be selected. This pauses the run until the **Go** option is reselected. When the run is paused icons within the Facility Window can be double-clicked to reveal their editor windows. One of the options within these editors is **Status** which reveals what the Resource is doing at the pause point, that is, is it busy, idle, broken down etc. This can be useful for debugging or investigating bottle-neck areas.

Select **Simulate ; Go** from the Alternative action bar

If you have done everything correctly you should observe your first AIM Alternative in action. In particular you will have observed the Resource icons change colour dynamically from blue (idle) to green (busy) and possibly also to purple (blocked).

Editing can be carried out during a pause period, but there are many changes to the Alternative and its parameters at this stage which will make it impossible to resume the run. It is then necessary to select either **Reinitialize** or **Restart** from the **Simulate** pull-down menu.

4.2 Data Output

Two forms of informative output are available - Dynamic Graphs which as the name implies are generated in the course of run, and Graphs/Reports after the event. The former require that a request is made *before* the run, and so the second category of graph/report is covered first.

On the **Facility** window, click on mach1 (be careful to click on the machine itself and not on any part which may be within it), hold down the Ctrl key and, while keeping it pressed, click on mach2 and mach3. From the Alternative window action bar select **Selected ; Graph**.

A further pull-down list appears which is in three sections, the first two of which correspond to **Utilization** and **Time in State**. The third section gives options for Dynamic Graphs and so is not available for selection at this point.

Select **Resource on Shift Utilization**. A graph of utilizations for the three machines should appear in an appropriately labelled window.

Close this window by clicking top left and selecting **Close**.. Repeat the above, only this time selecting **Resource Total Time in State (Bar Chart)**.

Close this window and repeat once more using **Resource Total Time in State (Pie Chart)**. This time there are three windows, one per bar chart. Close each of these windows.

Select **Simulate ; Restart** from the Alternative action bar.

Now select **Selected ; Reports**. Only one option is available, namely **Alternative Summary**. Select this and review it. Often the most immediately interesting figure in this report is the one at the rightmost end, so use the scroll bar to move to the right until the heading Production Rate appears, that is, the number of Parts which are produced per hour.

Close the window.

4.2.1 Dynamic Graphs

A Dynamic Graph is an alternative to animation, and as a run proceeds displays the ongoing value of, for example, a queue length or Resource Group utilization. Naturally the quantity to be displayed must be selected *before* the run commences.

On the **Facility window** click on mach1, hold down the Ctrl key and, while keeping it pressed, click on mach2 and mach3. From the Alternative window action bar select **Selected ; Graph**. From the third section select **Resource Queue Length**.

Select **Simulate ; Restart** from the Alternative action bar. You should now observe a Dynamic Graph of this quantity for each of the three machines.

A number of options are available to improve or vary dynamic graphs.

From the Alternative action bar do:

Options ; Y-axis ; High Value: 10 ; Push **OK**

Options ; Animate..

Use slider to change time between refreshes to 10 min.

Push **OK**

Simulate ; Restart

4.2.2 Reports

The following reports are available:

Alternative Summary
Resource Summary Resource Group Summary

The last two require that the Resource or Resources be selected by clicking on the appropriate icons in the Facility Window.

4.2.3 Graphs

The following graphs are available. Items in the second section require objects to be selected from the Facility window. Items in the third section are available for Dynamic Graphs only.

Resource On Shift Utilization
Resource Total Time in State (Bar Chart) Resource Total Time in State (Pie Chart) Resource Group on Shift Utilization Resource Group Total Time in State (Bar Chart) Resource Group Total Time in State (Pie Chart)
Resource State Resource Queue Length Resource Group Utilization Resource Group Queue Length

4.2.4 Status Reports

A further option is to obtain the status of a Load when a run is stopped.
Click on the icon representing the Load with the *right hand* mouse button.
This causes the following information to be displayed:

Order to which Load belongs

Current Jobstep

Resources, Pools and Material Handling Devices allocated

Resources currently needed

Values of any Attributes

Processing time at Jobstep, both total and remaining

Total processing and waiting time

It is also possible to edit the Load icon at this point by using a further
pull-down menu.

4.3 Saving Work

There are in general three ways in which work may be preserved, namely
by **Build/Save** which has already been discussed, by **Load/Unload**, and by
Restore/Back-up. Build/Save is done at the level of Alternatives,
Restore/Backup at the level of Database, Load/Unload can be done at
either.

To Load/Unload at the level of an Alternative use **Utilities** from the FACTOR/AIM action bar having first ensured that the Facility Window is closed by using **File ; Exit**. To Load/Unload at the Database level use **Database** from the FACTOR/AIM action bar. For Unloading Alternatives the default filename is **ALT00n.dmp** where n is the number of the Alternative. For Databases the default name is the name of the Database, again with extension **dmp**. The .dmp files provide the main means for conveying Alternatives from one AIM system to another.

Back-up/Restore is more comprehensive but also more time-consuming than Load/Unload since output data can be included, and also **User Files**, that is, C code which advanced users can incorporate to enrich an AIM model. Backup may not be performed on the same disk as that on which the Database itself resides. The general intention is that Backup should be onto another medium, for example 3.5 inch diskettes.

In the course of completing the activities in the boxes many default values were chosen which means that many alternative paths and opportunities were not explored. The next section revisits the Base case model and considers the possibilities offered by of some of these options.

4.4 Review of Base Case

From now on more condensed instructions for using AIM will be employed. These should help the reader who does not have an AIM system to understand the details both of components and processes which have to be conveyed to *any* simulation system which adopts a process-driven approach. The left hand column indicates where to navigate from, e.g. a window or action bar, and the remainder of each line indicates distinct actions separated by semi-colons. As before **bold** type indicates something to be looked out for on the screen. If it is the name of a graphical object, double-click on it. If it is the name of a push-button then push it, if it is the name of a radio-button select it, if it is the name of a check-box then check it (if it is to be unchecked, the word "uncheck" will be stated explicitly). In the case of a dialog box, a colon is used to separate the name of the field and the characters which have to be typed in it. If an item in bold type is not present on the screen it means it is a selectable item from a list-box. If there is more than one list-box on the screen the symbol " = " may be used to indicate the particular list-box in which the object to be selected is to be found.

Because there are a very large number of windows in AIM, it will not always be the case that all the windows in a sequence of actions will be named explicitly in the left hand column. Here is the first block of summary instructions to recreate Base case:

FACTOR/AIM	**Database** ; **New** (or **Open** if database exists)
action-bar	**Database Name:** Trial ; **OK**
	Simulate ; **Build/Run**
	Description: Base case ; **Build**
Simulator action-bar	**Options** ; **Build..** ; **Extended** ; **OK**
	Edit ; **Add..**
	drag **Add Components** window to top right
Facility/Add Comps	Add and position 3 m/cs.
	Add and position 1 multi-capacity m/c.
	Add and position 1 Wip pool.
	Add and position 1 conveyor segment to join

	mach1 and mmac1.
	Add and position 1 Time-Persistent Value.
Add Components	**Variable ; Create ; Integer ; OK**
Facility	**tpval1 ; Expr**: var1 ; **OK**
	Desc: tput = ; **OK**
Add Components	**Part ; Create ; OK ; Part ; Create ; OK**
	Demand ; Create
Demand Editor	**Number of Parts to Release**: 4
	Inter-arrival Time: exponential(1)
	Part: part1 ; **OK**
Add Components	**Demand ; Create**
Demand Editor	**Number of Parts to Release**: 6
	Inter-arrival Time: exponential(1.25)
	Part: part2 ; **OK**

The next section of the summary instructions relates to the construction of the Process Plan which in general takes up as much time as all the other parts of the Alternative put together.

Add Components	**Process Plan ; Create ; Insert ; OK**
Set-up Jobstep Editor	**Desc**: Pull parts into buffer
	Name = wip1 ; Units = 1 ; Action = Allocate ; OK
	Insert ; OK
Set-up Jobstep Editor	**Desc**: Make
	Name = wip1 ; Units = 1 ; Action = Free After
	Name = mach1; Units = 1 ; Action = Allocate ; OK

74

	Insert
	Operation Time = normal(.1,.02) ; **OK** ; **Insert**
	Type = **Move-between** ; **OK**
Move-between Editor	**Desc:** Convey from make to test
	Conveyor ; System = **csys1 ; Origin** = **ccp1**
	Destination = **ccp2**
	Free at Pickup = **mach1**
	Allocate at Drop-off = **mmac1** ; **OK** ; **Insert** ; **OK**
Set-up Jobstep Editor	**Desc:** Test
	Name = **mmac1 ; Units** = **1 ; Action** = **Free End**
	Operation Time = uniform(.2,.3) ; **OK** ; **Insert**
	Type = **Probabilistic Select** ; **OK**
Prob Select js Editor	**Desc:** Decision following test
	Jobstep = **js1** ; **OK** ; **Insert** ; **OK**
Set-up Jobstep Editor	**Desc:** Rework following test
	Name = **mach2 ; Units** = **1 ; Action** = **Alloc/Free**
	Operation Time = normal(.5,.2) ; **OK** ; **Insert** ; **OK**
Set-up Jobstep Editor	**Desc:** Final assembly
	Name = **mach3 ; Units** = **1 ; Action** = **Alloc/Free**
	Operation Time = .1 ; **OK** ; **Insert** ; **OK**
	Type = **Assign** ; **OK**
	Name = **var1 ; Expr:** var1 + 1 ; **OK**
Process Plan Editor	**js5**
Prob Select js Editor	**Jobstep** = **js6 ; Probability** = **.1** ; **OK** ; **New**
	Jobstep = **js7 ; Probability** = **.9** ; **OK** ; **OK**
Process Plan Editor	**Done ; Edit ; Find/Select**
	Type = **Demand ; Search ; dord1**
Demand editor	**Details.. ; Process Plan** = **pplan1**
	Expected Makespan: .5 ; **OK** ; **OK** ; **dord2**

75

	Details.. ; Process Plan = pplan1
	Expected Makespan: .5 ; OK ; OK
Process Plan Editor	**Done**
Add Components	**Close**

The final part of the summary instructions refers to the check, run and view output stages.

Alternative action-bar	**Utilities ; Check Model**
	Options ; Simulate..
	Start Time: 0 ; Finish Time: 3 ; OK
	Simulate ; Go
	mach1, mach2, mach3
	(keep Ctrl key depressed)
	Selected ; Graph ; Time in State (Bar Chart)
	Close Selected
	Reports ; Alternative Summary
	File ; Save ; Save

With practice the activities in the last three boxes could be accomplished from scratch within about twenty minutes.

The next section is about adding enhancements to the Base case. This is intended to give more practice in navigating through AIM, thereby increasing familiarity with the sequences of windows which commonly occur.

4.5 Enhancements to Base case

During animation the state of the Machines is indicated by colour codes applied to the Machine icons. However, no change takes place in the Multi-capacity Machine icons because at any given time each of these objects will be partially busy and partially idle. It is possible, however, to indicate the overall busyness by displaying the number of units currently free. The next box shows how to do this and at the same time remove some of the verbal clutter from the Facility Window.

FACTOR/AIM	**Base case ; Build**
Simulator action-bar	**Options ; Build**
	Uncheck **Show Control Points**
	Uncheck **Show Segments**
Facility	**mmac1**
	Animation Style ; Icon and Count ; OK
	Simulate ; Go

A variety of icons is available to represent parts – the next box shows how to access and use these.

```
Simulator action-bar    Edit ; Find/Select ; Part ; part1 ; Icon..
                        Early ; PAR001EA ; Late ; PAR001LA ; OK
                        part2 ; Icon..
                        Early ; PAR002EA ; Late ; PAR002LA ; OK
                        Simulate ; Go
```

To check that the number of parts through rework is indeed around 10% to correspond to the probability of 0.1 in js5, add more Time-Persistent Values for the other machines.

Just as time is determined in units of hours, so the primary unit of length in the Facility Window is determined in feet. The scale of the Facility Window can be adjusted by changing the Default Segment Length/Pixel from **Options ; Build** on the simulator action bar. The scale of the Facility Window can be displayed explicitly by using the **grid** option (see section 1.9.7). Conveyors have a constant velocity determined by the combination of time (specified in the Move-Between Jobstep), length (as defined in the Facility Window) and capacity (as defined in the Conveyor Segment Editor), that is, the number of loads which the conveyor can accommodate when fully populated.

The model so far has no **Operators**, nor are there any **Setup times**. In the most straightforward case a Setup time is a penalty involved when a machine changes from running one part to running another. Incorporating Operators and Setups is illustrated in the next box in which two Operators are added, one to do Setups on the first two Machines, and the other to mind the third machine permanently. A Setup involves two objects, a **Resource** to be setup, typically a machine, and a **Required** Resource, typically an Operator. A setup also requires a **Rule**, for which the default is that a setup takes place on the **Basis** of a change of Part-number (see section 5.3 for a further discussion on the possible criteria for Setup).

Simulator action-bar	**Edit ; Add..**
	Add two Operators by dragging and dropping
	into suitable positions in the Facility Window.
	Edit ; Find/Select..
	Select **Process Plan** ; edit **pplan1**
	Double-click on **js2**.
Jobstep Editor	**Setup ; Resource = mach1 ; Required = oper1**
	Setup Time ; Expr = 0.15 ; **OK**
	Repeat for **js6** with mach2 ; Double-click on **js7**.
	Resource/Pool/Group
	Name = oper2 ; Units = 1
	Action = Alloc/free ; OK ; Done
	Simulate ; Restart
	Obtain a barchart of Time in State for the two
	operators.

If a Setup is required for every incoming Load, then the Rule should be changed to **Always**.

Edit **js2** and **js6** by making the following changes:

Jobstep Editor	**Setup ; Rule = Always**
	(The **Basis** field is ignored in this case)
	Simulate ; Restart
	When mach1 is in setup state it is shown in dark green.
	Obtain a barchart of Time in State for the two operators and compare the busyness of oper1 with the previous run.

Next some inventory items are added.   There are three types of entity involved, a graphical item on the Facility Window called a **Material**, and two Jobstep types, **Add-to-Material** and **Remove-from-Material**.   In the ongoing example, the Materials might be three components which are added to a kit at the start of the manufacturing sequence, and are eventually built into the finished product at Final Assembly. To model wastage it is assumed that four components are removed at this stage, so that it is only the presence of incoming WIP which sustains the stock of Material.

Alternative action bar	Add a **Material** by dragging and dropping it to a position near mach1.
Material Editor	Change **Capacity** to 100 and **Initial Level** to 0. Select radio button **Count** (default = **Tank**)
Process Plan Editor	Click on **js1** ; **Insert** ; **Add-to-Material (js9)**
Jobstep Editor	Select **matl1** ; **Quantity** 3
	Procedure: Part ; OK
Process Plan Editor	Click on **js7** ; **Insert**
	Remove-from-Material (js10)
Jobstep Editor	Select **matl1** ; **Quantity** 4
	Procedure: Part ; OK ; Done
	Simulate ; Restart

Add-to-Material can also appear towards the *end* of a Process Plan to model dispatch to a shipping area, and conversely Remove-from-Material often appears towards the start to model kits being assembled by making withdrawals from storage bins.

At this stage the Facility window might look something like this:

81

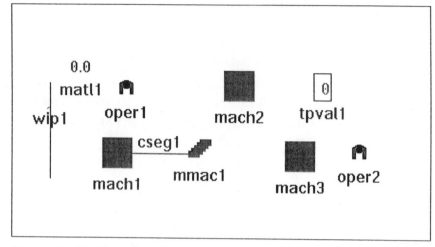

Figure 12. Facility Window with Operators and Material

4.6 Breakdowns and Maintenance

No real manufacturing plant operates without **Breakdowns** and **Maintenance**. The pattern of the former is usually only known within the limits of statistical distributions, while that of the latter may vary at the one extreme from a regular and rigid programme to that of a Breakdown type of distribution at the other. Breakdowns and Maintenance are perceived as non-graphical entities with their own editors and in the course of editing they are attached to particular Resources. Data for Breakdowns is typically obtained from observation on the ground or from past records, and this is another point at which empirical fitting of probability distributions together with the use of a standard statistics package is appropriate (see section 3.4).

Both Breakdowns and Maintenance can be triggered in four essentially different ways, viz:

by a count of the number of Loads processed;

by total elapsed time;

by total on-shift time;

by elapsed processing time.

4.6.1 MTTF and MTR

The quantities in the last box which were called "Value Between Breakdowns" and "Repair Time" are frequently referred to as MTTF (Mean Time Between Failures) and MTR (Mean Time to Repair) respectively. Sometimes the latter is denoted by MTTR. Manufacturing lines can often have quite different sensitivities to these two quantities, and an important area of simulation from the application point of view is in advising management whether engineering effort is better directed to improving reliability, that is, increasing MTTF, or to shortening repair times. There are no general rules but case studies exist in which simulation demonstrated that changing the design of tools so that they could be removed and replaced by line operators rather than by maintenance engineers was much more fruitful in increasing production than attempts to improve machine reliability, that is, in such cases reducing MTR is more effective than increasing MTTF.

4.7 Shifts and Shift Exceptions

In common parlance the word "shift" is used in two senses, first to denote a particular time-period in the usage "something went wrong on the early shift," and secondly in the sense of the collection of people who regularly work together in a shared work pattern as in "the night shift were responsible for the breakdown." A **Shift** as used in AIM veers towards the second of these meanings. It is the name given to a work pattern of up to seven days which is shared by a collection of Resources, Material Handling Devices and Demands which come together for the purposes of production. In the absence of any attached Shifts, Resources and Material Handling Devices are available and Orders will flow for 24 hours per day and seven days per week.

Several Resources or Demands may share the same Shift. It is also possible to define **Shift Exceptions**, that is, periods during which either the regular pattern associated with a Shift is suspended, or an extra Shift is brought into operation. Public holidays and absenteeism/strikes are an example of the first type of Shift Exception, while extra overtime to meet peaks of demand is an example of the second.

4.8 Lookup Tables

Operation times are often Part-dependent, and matrices called **Lookup Tables** are a convenient way of recording values which require two quantities to define them, e.g. a Part name and a Jobstep name. For example:

	js2	js6	js7
part1	0.1	0.5	0.1
part2	0.2	0.4	0.2

Lookup tables have the effect of separating the *numerical* parameters from the *structural* aspects of an Alternative. They also allows multiple parameter changes to be made in a single edit activity.

Add Comps window	Create a **Lookup Table**
Lookup table Editor	Insert part1 and part2 as row headers in the leftmost column.
	Insert js2, js6 and js7 as column headers in the top row.
	Fill in first row values as 0.1, 0.5, 0.1
	Fill in second row values as 0.2, 0.4, 0.2 ; **OK**
Process Plan Editor	Click on **js2**
Jobstep Editor	Replace normal(0.1,.02) in **Operation Time** with normal(lookup("ltbl1" ,partname,jsname),0.02).
	Make the same replacement in **js6** and **js7**.
	Simulate ; Restart

Up to this point Alternatives have been discussed in relative isolation (the illustrative example in the boxes has so far used only one). However, as the name implies, the essential idea is that in a substantial Project a number of *different* Alternatives should be constructed in order to compare different structural possibilities, whereas different parameter values within the same structure are provided for by Lookup Tables. One of the strengths of the AIM product is the ease with which different Alternatives can be compared through graphs and reports.

In the Base Case example, parts from Rework are passed immediately to Final Assembly. It might be interesting to observe the differences to overall performance which would follow from making reworked parts revisit the Test machine. This is achieved by creating a second Alternative as follows:

FACTOR/AIM	**Database: TRIAL**
	Click on Base Case ; **Utilities ; Copy**
	Name = Second shot
Alternatives window	Click on Second shot ; **Simulate ; Build/Run**
Process Plan Editor	Click on **js6**
Jobstep Editor	**Next** = **js4** (replacing **js7**) ; **OK**
Process Plan Editor	Click on **js4**
Jobstep Editor	**Action** = **Alloc/Free ; OK**
Process Plan Editor	Click on **js3**
Jobstep Editor	Clear **Allocate at Drop-off** field (was **mmac1**)
	OK ; Done ; File ; Save
Alternatives window	Hold Ctrl key and click on both Alternatives.
FACTOR/AIM	**Simulate ; Batch Simulate**

	Finish Time = 10 ; **Number of Runs** = 2
	Do the same for the second Alternative.
Batch simulator	Observe two Replicates run for each of the two
	Alternatives.
FACTOR/AIM	**Evaluate ; Graph ; Category ; Resource/Group**
	OK
	Result is On-shift Utilization Barchart with
	the results of the two Alternatives compared
	in side-by-side bars.

The following diagram shows the default style of a typical side-by-side barchart to compare utilizations in two Alternatives. This and other charts like it are obtainable with virtually no further effort following a run, and are invaluable as the basis for Project reports.

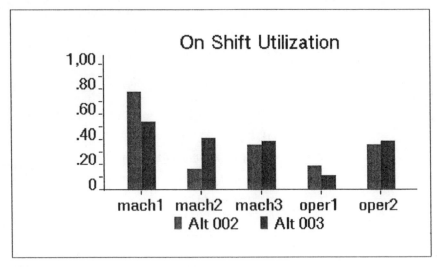

Figure 13. Specimen AIM graphics

4.9 Building an Alternative from scratch

In the previous box **Copy** was used to construct a new Alternative. Sometimes it is easier to start with an empty Alternative. From the Alternatives
Window, instead of clicking on one of the existing Alternatives hold the
Alternate key and click to deselect any which are already highlighted. Then
do **Simulate ; Build/Run** as in the third line of the immediately previous
boxed section, and proceed with an empty Alternative.

5. REFINEMENTS
OF BASIC CONCEPTS

5.1 Pulls

Pulls were defined in section 1.4 as an alternative mechanism to Demands for releasing Orders into an Alternative. Pulls are driven by the level of a single Material, and one (and only one) Part and one (and only one) Material are referenced by a Pull. There are three **Pull Release Rules**:

1. **Part Removal** - release an Order every time a unit is removed from a Material;

2. **Quantity** - the Pull is activated by a withdrawal from Material which also brings its level below the **Release Value** level. In inventory control literature this is called the **Reorder Level** - see Ref. [8];

3. **Need** - like Quantity, but taking into account Material needed for outstanding Orders. Since some of these may be queued, and thus Material needed for them not yet withdrawn, this sort of Pull release may be triggered by an event other than a Material withdrawal.

The illustrative example continues with an amendment to the "Second Shot" Alternative given in section 4.8. The restated problem is:

Parts with a single Part number arrive in a buffer from which they are **pulled** into the first machine whose operation is known as making. They then move by conveyor to a series of identical test machines. Those which are accepted go direct to a final assembly machine, otherwise they go to a rework machine following which they are retested. A scoreboard after final assembly records the total numbers which have passed through the system.

The Alternative is to be reconstructed so that **part1**'s are fed into the system using Pulls, and the Demand for **part2** is removed.

Add Comps window	Create a **Pull**
Pull Editor	**Number of Parts to Release:** 10
	Release Rule: Need ; Release Quantity : 5
	Material : mat11 ; Part : part1 ; Details..
	Process Plan: pplan1 ; Makespan: 0.5
Simulator Action-bar	**Edit ; Find/Select.. ; dord1**
Demand Editor	**Inter-arrival Time :**INFINITY

```
Simulator Action-bar    Edit ; Find/Select.. ; dord2
Simulator Action-bar    Edit ; Clear


                        Simulate ; Restart
```

There is an essential distinction between Parts, Demands and Pulls on the one hand, and Orders, Loads and Batches on the other. The former are *static* entities in the sense that they exist and can be described and discussed independently of whether or not any run has taken place. The latter are *dynamic*, that is, they only come into existence after an Alternative has started to run. Dynamic entities thus have the property of possessing **Status**, for example a Load can be **In-process, Waiting** or **Blocked**.

Orders are **released** by Demand and Pulls. There are two further ways in which Orders can be instantiated. The first of these is by Creating **In-Process** Orders, that is, Orders which exist at the start of a run, and therefore possess Status at this point. The second way is by means of a **Release** Jobstep, which is provided to deal with the sort of situation where a pallet parts company with its Load and continues its separate existence (e.g. by returning to pallet stock) via its own Process Plan. There are thus three types of Order :

1. New - the standard type generated by Demands and Pulls;
2. In-Process - existing at the start of the run;
3. Explicit-Release - see next box for how this is used.

In the "Second Shot" example suppose that Parts arrive at the "Make" machine on pallets from which they become detached at the end of the Operation. The pallets themselves go through a further process of some sort

91

which returns them into circulation again. The following box describes how this would be modelled in AIM.

Create a new Part and give it the name pallet.

Create a new integer Variable (var2) to count the number of pallets.

Add **mach4** to the Facility Window to represent what happens to the pallets on their return journey.

Create a new Process Plan (pplan2) and insert two Jobsteps, namely a Setup/Operation to allocate/free **mach4** with Operation time 0.1, and an Assign to var2 whose Expression is var2 + 1

Create an Order (order1). Make it Explicit-Release using the radio button and associate it with Part pallet and Process Plan pplan2.

Insert a Release Jobstep after js2 in pplan1 and give it order1. Create a Time-Persistent Value to display var2.

Run the simulation.

Where a Demand or Pull prescribes multiple Parts per Load and also multiple Parts per Order, it may be that Orders do not divide evenly into

Loads. **Excess Options** are necessary to deal with the possibilities which might be required, viz:

Add to last Load;
Use as a new Load;
Supplement with extra Parts.

Another manufacturing entity which is associated with Parts is a **Fixture**, for example a jig or work-board holder which remains with the Part for all or part of the time that it is in the course of manufacture. Unlike a pallet in the illustration above, a Fixture does not participate in a process on its own account. It is instead a particular type of Resource which characteristically remains allocated to a Part through several steps of the Process Plan.

A **General Resource** is like a Fixture except that it does not necessarily remain with its Part while it is in a Pool (see section 1.5).

5.2 Just-in-Time Systems

JIT systems are characterised by the association of Loads with authorisation **Tags** which are commonly called kanbans. JIT systems are Pull systems which minimize inventory by requiring Operations not only to have a Load present but also to have its appropriate Tag before the Operation can begin. Resources in a kanban-driven system are no different from those in a Demand or Pull system. However in the Process Plan each "making" step is a **Kanban Jobstep** rather than an Operation Jobstep.

Tags work like the tablets used to ensure safety in a single-line railway. When a Load leaves a Resource it is accompanied by a Tag which is not freed until another Operation begins at the next Resource. The Tags themselves are modelled either as WIPs or as General Pools. The number of units in the WIP or General Pool is the maximum inventory level (often one) between these particular Resources. The distinction between a WIP and a General Pool is that each space in a WIP must be allocated and freed on behalf of the same Load, whereas this need not be the case for a General Pool, in which space can become available as the result of a different Load being freed. In most cases the practical distinction in AIM between using a WIP and using a General Pool is that the former leads to an obvious animation of the inventory buffer, whereas the latter produces graphs and reports of its utilization.

The Kanban Jobstep provides a very simple way of modelling sequences of Operations where the sort of orderly discipline described above applies. In particular it avoids much of the need for explicit allocation and freeing of Resources.

5.2.1 A Kanban system modelled in AIM

This section contains instructions to model a simple Kanban system in AIM. Suppose that there are three machines in sequence connected by Conveyors, and the Operation times are in increasing sequence so that blocking would occur in the absence of Kanban Tags. The Facility Window for the configuration is as shown:

94

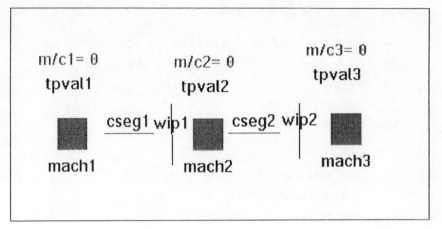

Figure 14. Kanban system

In the Process Plan, Jobsteps which would normally be of type Operation or Setup/Operation are replaced by Kanban Jobsteps.

Create the objects shown in the above diagram, and give the wips a capacity of 4.

Create three integer variables to record the count of Parts processed by the three machines, and create three corresponding Time-Persistent Values to display them as shown in the diagram.

Create two Conveyor Segments to connect the Machines as shown.

Create a Part (part1).

95

Create a Demand which releases 15 Parts at a time, one Part per Load with an Inter-arrival Time of 1, so that 15 Parts arrive in a burst every hour.

Define a Process Plan with the following Jobsteps:

Kanban
  Operation Time = 0.01667 (1 minute)
  Kanban Machine = mach1
  Required Tag / Out buffer = wip1
Assign (i.e. count variable for mach1)
Move-Between
Kanban
Operation Time = 0.025 (1.5 minutes)
  Release Tag / In buffer = wip1
  Kanban Machine = mach2
  Required Tag / Out buffer = wip2
Assign (i.e. count variable for mach2)
Move-Between
Kanban
  Operation Time = 0.0333 (2 minutes)
  Release Tag/ In buffer = wip2
  Kanban Machine = mach3
Assign (i.e. count variable for mach3)

Run the simulation. You should observe a Kanban system with four Tags.

Now change the **Units** fields in each of the Kanban Jobsteps to 4.

> Rerun the simulation. You should now observe a Kanban system
> with one Tag.

5.3 Setups

In section 4.5 a Setup time was defined as a penalty involved when a
machine changes from running one part to running another. A Setup
involves two objects, the **Resource** to be setup, and a **Required Resource** to
carry out the Setup. It is also necessary to specify the *criterion* for incurring
a Setup penalty. Parts can be thought of in terms of a **Hierarchy** of Family,
Subfamily, and Part-number. Possible criteria are that a Setup may be
required for each new Load, or for changes in Family, Subfamily, Part
number or Order.

5.4 Generalised Selection

The "Base Case" example in Chapters 1 and 2 used Probabilistic Selection
as a basis for making a branching decision. More general decisions can be
made using the **Select Jobstep**. At this point the AIM user comes closest to
conventional programming. Conditions might involve the values of a Variable or Attribute, or the Partname of the Part arriving at the Jobstep. In all
cases the objective of the Select Jobstep is to resolve the Next Jobstep. If
the Condition Rule of the Select Jobstep is selected, an ordered list of conditions must also be supplied, each of which references one and only one

97

Jobstep. The list is scanned in order and the Next Jobstep is determined by the first condition on the list which is found to be true.

Sometimes there is a set of Jobsteps from which the Next Jobstep is selected in cyclic order, that is, after every selection the next one in cyclic order is marked as Next Jobstep for the next Load arriving at the Select Jobstep. The Rule in this case is **Cyclic**. A variation on this Rule is **Cyclic/Resource**, which means that if the first Resource specified to be either freed or allocated in the designated Next Jobstep is not available, then the next one in the cycle is to be chosen. If none of the cycle of Jobsteps has its first Resource available then the Next Jobstep is the one which would be determined by the Cycle Rule.

Frequently, routing decisions are made on the basis of availability of Resources, for example when a Load should be directed to the first available free Operator. This case is described by one of two search Rules **Search/First** and **Search/Last**. As with the Condition Rule, a list of possible Next Jobsteps is designated, and the one chosen dynamically is the first/last for which the first named Resource or Resource Group in its Allocation List is available. Every Jobstep in the list must have at least one Resource or Resource Group in its Allocation List.

Five Rules and three Rule types have thus been identified:

Condition;
Search First/Last;
Cyclic with and without Resource.

5.5 Inspection

The **Inspect Jobstep** is a refinement of the Probabilistic Select Jobstep. It facilitates the modelling of inspection processes and contains an Operation time followed by rerouting and possibly scrapping of Parts.

If inspection is at the level of a Load, that is, each Load is deemed to pass or fail with subsequent rerouting, then the Inspect Jobstep is just like an Operation Jobstep followed in sequence by a Probabilistic Jobstep. Another possibility is that an average *proportion* of Parts in every Load fail, the number failing in any particular Load being determined by random number generation. The failed parts form a new Load which is routed appropriately, possibly to a Jobstep which scraps them.

6. FORMULAE FROM QUEUEING THEORY

6.1 Analytic Methods

Before proceeding to consider those aspects of manufacturing processes which make their simulation a *complex* matter it is appropriate to consider the extent to which *simple* results can be obtained by purely analytic methods, that is without any recourse to simulation at all. This is the point at which the simulation analyst should be prepared to invoke some of the elementary results of Queueing Theory.

6.2 Kendall's Notation

In the early 1950's Professor D.G.Kendall of Cambridge University formulated a classification system to describe single queues,[*] that is, queues which customers join in a random pattern, and leave after obtaining a single service, the times for which are also distributed according to a random pattern. This notation has obtained widespread acceptance in Operational Research and is used in the summary of results from elementary Queueing Theory which form the basis of this Chapter. The form of a queue is described by up to six parameters, viz.

1. inter-arrival time distribution for which some of the codes are:
   M = exponential
   G = general (that is, exact distribution is not specified)
   D = deterministic (that is, constant)
   E_k = Erlang k (see section 3.2);
2. service time distribution;
3. number of servers;
4. system capacity;
5. number in source;
6. queue discipline.

In practice it is rare for more than the first three of these parameters to be used so typical queues are M/M/1, that is a single server queue for which both service time and inter-arrival time are exponentially distributed, and M/G/1 which is the same except that the service time has a general distribution.

The input parameters for a queue are

λ = mean inter-arrival rate;

---

[*] see Kendall D.G, 1953, "Stochastic Processes Occurring in the Theory of Queues and their Analysis by the Method of the imbedded Markov Chain," Ann. Math. Stat., 338-354.

101

μ = mean service rate;

c = number of servers.

The average inter-arrival and service *times* are the reciprocals of the corresponding rates. A steady state requires that $\lambda \le \mu$, otherwise the service point must in the long run become overloaded, and the length of the queue increase without limit.

There are two derived parameters namely:

traffic intensity $u = \dfrac{\lambda}{\mu}$;

server utilization $\rho = \dfrac{\lambda}{c\mu}$.

A Part flowing through an M/M/1 system can be in one of two states, viz. queueing and service. The times spent in those states are denoted by q and s respectively, and the total time in transit is denoted by t so that

$$t = q + s$$

The symbol N subscripted by q, s or t denotes the numbers in the various states, and the expression E[x] denotes the *Expected Value* of x which is the statistician's way of saying "average" in the case where there is a potentially infinite number of values of x. In a steady state Parts must be either queueing or in service and so it is the case that

$$N_t = N_q + N_s$$

and also

$$E[N_q] = \lambda E[q] \qquad E[N_s] = \lambda E[s] \qquad E[N_t] = \lambda E[t]$$

102

λ is the multiplier in all three of the above equations, which expresses the fact that in a steady state it is the *arrival* rate which determines both service and overall performance.

The third of these formulae is often used as a rule of thumb, and is known as Little's Law[*]. It can be mnemonically rendered as

$$L = IT$$

where

L = average numbers in transit ($E[N_t]$);

I = average inter-arrival rate (λ);

T = average time spent in system ($E[t]$).

Little's Law holds for a very wide range of queueing systems. It is an equation which describes the *conservation* of queue length in a steady state. For example, assume that in a steady state an average queue length of two is observed and on average it takes customers 20 minutes to pass through the system. According to Little's Law, $\lambda = \frac{1}{10}$, that is the average inter-arrival time is ten minutes. Reducing this time to 20 minutes would halve the average queue length. This kind of "broad brush" style of calculation is typical of the elementary uses of Queueing Theory which should be the basis of ancillary sanity checks in the course of simulation analysis.

In calculating average times for queueing, two types of average are of interest, namely averages over all Parts, and averages over only those Parts

---

[*] see Little J.D.C., 1961, "A Proof of the Queueing Formula L = λW" , Opns. Res., 9, 383-387

which queued, that is ignoring those which went through without any queueing. Symbolically this is q | q > 0.

6.3 M/M/1 Queues

The M/M/1 queue is very tractable analytically and has the following properties:

Times	Mean	Variance	pth.Percentile
t	$\dfrac{1}{\mu(1-\rho)}$	$\dfrac{1}{\mu^2(1-\rho)^2}$	$E[t]\log\dfrac{100}{100-p}$
s	$\dfrac{1}{\mu}$	$\dfrac{1}{\mu^2}$	$E[s]\log\dfrac{100}{100-p}$
q	$\dfrac{\rho}{\mu(1-\rho)}$	$\dfrac{(2-\rho)\rho}{\mu^2(1-\rho)^2}$	$E[t]\log\dfrac{100\rho}{100-p}$
q|q > 0	$\dfrac{1}{\mu(1-\rho)}$	$\dfrac{1}{\mu^2(1-\rho)^2}$	

The values in the percentile column give some useful rules of thumb, e.g. the 90th and 95th percentiles for t are at approximately 2.3 and 3 times the mean value of t.

Numbers

	Mean	Variance
N_t	$\dfrac{\rho}{1-\rho}$	$\dfrac{\rho}{(1-\rho)^2}$
N_s	ρ	$\rho(1+\rho)$
N_q	$\dfrac{\rho^2}{1-\rho}$	$\dfrac{\rho^2(1+\rho-\rho^2)}{(1-\rho)^2}$
$N_{q|q>0}$	$\dfrac{1}{1-\rho}$	$\dfrac{\rho}{(1-\rho)^2}$

Two simple graphs convey the essence of M/M/1 queue behaviour, namely those of the ratios of transit time and queueing time to service time when plotted against utilization, that is

$$\frac{E[t]}{E[s]} \text{ vs. } \rho \text{ and } \frac{E[q]}{E[s]} \text{ vs. } \rho.$$

Both graphs deliver the same message, namely that, under exponential conditions, when ρ increases to a value between 0.6 and 0.7 the service level perceived by the customer rapidly deteriorates. This is further emphasised in section 6.7.

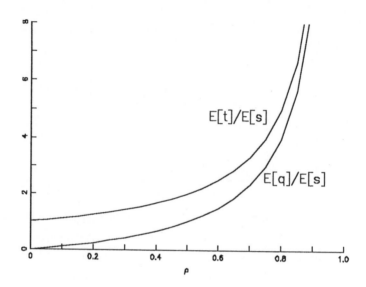

Figure 15. Graphs of Transit-time and Queueing-time ratios

6.4 M/M/1/k Queues

An M/M/1 queue which has a finite capacity k is described as an M/M/1/k queue. This means that a customer who arrives and finds k items already in the queue is **balked**, that is, exits the system immediately. If further, $\lambda = \mu$ then

$$E[N_t] = 0.5k$$

If λ and μ are not equal

$$E[N_t] = \frac{u(1 - (k+1)u^k + ku^{k+1})}{(1-u)(1-u^{k+1})}$$

where $u = \frac{\lambda}{\mu}$. In both cases the average transit time is

$$\frac{E[N_t]}{\lambda_a}$$

where λ_a is the *actual* rate at which customers enter the system. This is greater than the *true* rate on account of those Parts which are turned away because of the capacity limitation. Specifically if $\lambda \neq \mu$:

$$\lambda_a = \frac{\lambda(1 - u^k)}{1 - u^{k+1}}$$

an expression which tends to $\dfrac{\lambda k}{k+1}$ as $\lambda \to \mu$.

6.5 M/G/1 Queues

For the M/G/1 model of which the M/M/1 model is a special case the following formulae for Expected Values apply:

Numbers

N_q	$E[N_q] = \dfrac{\lambda^2 \sigma_s^2 + \rho^2}{2(1-\rho)}$
N_t	$E[N_q] + \rho$

Times

t	$\dfrac{E[N_t]}{\lambda}$
q	$\dfrac{E[N_q]}{\lambda}$
q|q > 0	$\dfrac{E[q]}{\rho}$

The first line in the above table is just Little's Law once again.

The M/D/1 model a special case of the M/G/1 model in which $\sigma_s^2 = 0$.

6.6 Coefficient of Variation

The **Coefficient of Variation** (**cov**) is an important statistic for making quick numerical assessments of queueing performance. It is defined as the ratio of the standard deviation to the mean. Provided the mean is not zero, which can usually be taken for granted for quantities like queue lengths and waiting times, the Coefficient of Variation provides a convenient "rule of thumb" measure for the scatteredness of a distribution. The exponential distribution has $cov = 1$ and this provides a general yardstick for variability. Some feel for the "variableness" of the exponential distribution can be gained from the illustration in section 6.7.

For both the Weibull and gamma distribution the Coefficient of Variation depends on the shape only. The gamma distribution has a particularly simple Coefficient of Variation, namely

$$cov = \frac{1}{\sqrt{\beta}}$$

which means that as β increases the distribution gets less and less variable. Thus gamma distributions with $\beta < 1$ one are in some sense "more variable" than exponential where "more variable" distributions are characterised by having relatively higher numbers of large values in their tail.

The numbers table for the M/G/1 queue in section 6.5 gives an indication of the sensitivity of queues to changes in the Coefficient of Variation. The first formula

$$E[N_q] = \frac{\lambda^2 \sigma_s^2 + \rho^2}{2(1 - \rho)}$$

may be rewritten

$$E[N_q] = \frac{\rho^2(1 + cov_s^2)}{2(1 - \rho)}$$

which shows that for constant ρ the numbers in the queue increase according to the *square* of the Coefficient of Variation.

6.7 Example of the use of analytical formulae

A plant has a rework area which operates one eight-hour shift per day. The mean time to repair a Part is 30 minutes, and the average number of Parts brought in for repair each day is ten. Ten times 30 minutes is five hours which suggests that there should be plenty repair capacity; however,

109

the manager concerned receives constant complaints about the delays in repairing Parts.

Assume an M/M/1 model and apply the formula for $E[q]$ above.

ρ (average utilization) = .625,

$\dfrac{1}{\mu}$ = 30 mins.

so $E[q]$ = 50 minutes with variance 5500 $mins^2$, that is, a standard deviation of about 75 mins. This gives a Coefficient of Variation of 1.5 which indicates the considerable variability of this statistic.

Also $E[N_q]$ is just over one, so putting these two calculations together there is likely to be substance in the complaints, in spite of the relative low overall utilization.

Refs. [1] and [10] contain good collections of analytic results for queues, together with admirably readable accounts of the mathematics involved.

6.8 Example of sensitivity to Coefficient of Variation

The graph below shows the results of a simulation experiment in which Parts were processed in turn by each of three machines in series. The inter-arrival time of incoming Parts was constant and could produce a throughput of 2,000 if the service time allowed. The simulated service time was in fact Normal with a fixed mean slightly less than the inter-arrival time. The objective of the experiment was to study the combined effects of Coefficient of Variation and Buffer Size on overall throughput. The results of 24 runs are shown in the graph below which shows clearly how sensitive throughput is to what are in effect quite small changes in standard deviation.

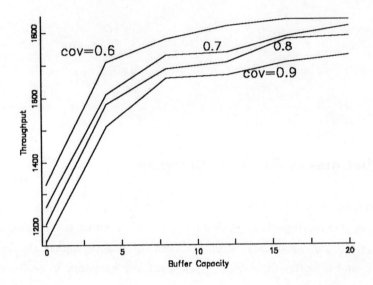

Figure 16. Throughput vs. Buffer for varying Coeff. of Variation

111

7. COMPLEXITY

7.1 What makes Factories Complex

This chapter is about the three principal characteristics of factories which make them complex organisations. It adds appropriate words and phrases to the manufacturing vocabulary which has been the ongoing thread through the book, and indicates the Jobstep types which are necessary to deal with them.

The three sources of complexity are;

Aggregation and Disaggregation;
Movement;
Sequencing, Selection and Contention.

7.2 Aggregation and Disaggregation

The primary aggregations are those of Parts, which in the *customer's* perception are aggregated into Orders, and in the *manufacturer's* perception are aggregated into Loads.

In almost every sort of factory there are stages of manufacturing in which aggregations of Parts change size in one way or another. An important part of constructing a generalized framework for analysing manufacturing processes is the classification of the different sorts of circumstances in which such aggregations and disaggregations occur.

7.2.1 Changing Load Size

Sometimes a Load may change its size because Parts are despatched to another activity outside the scope of their own Project, in which case the Load-size is decremented. It may of course happen that Parts are *received* from another Project, in which case Load-size is incremented. Another possibility is that a Load consumes some of its own Parts in the course of an Operation, which again would result in a reduction in Load-size. Yet another possibility is that the number of Parts is *multiplied* as a result of an Operation. All of these circumstances can be modelled by the **Change-Load-Size** Jobstep.

7.2.2 Produce and Release

The Release Jobstep has already been discussed in section 5.1. This produces new Loads as a result of releasing new Orders and is designed to deal with situations like pallet processing where something which has been integral with a Load for a while becomes detached and continues its existence using a separate Process Plan.

The Produce Jobstep can do one of two things, namely

1. create a new Load which can be directed to other Jobsteps in the Process Plan; or
2. add to one or more Materials.

In its second usage it is simply a multiple version of the Add-to-Material Jobstep. The former use is more interesting in that new Loads can be created and routed to an alternative Jobstep within the same Process Plan. This is relevant to situations where raw materials such as steel rolls enter a factory and are immediately cut up into parts of different shapes for distribution to different functions within it.

Optionally it is possible to kill off the first Load at the Produce Jobstep, so that only the Produced Load goes forward.

7.2.3 Assembly

Another form of disaggregation is observed in those manufacturing operations which begin with a kitting process in which several Materials are assembled to form a kit for presentation to subsequent Machines and Operators. This process is called Assembly and is described by the **Assemble Jobstep**. The distinctive features of Assembly are that Material is involved, and that transformation of Parts is brought about by reduction in Material.

114

The following simple Alternative illustrates the basic features of the Assemble Jobstep:

Create mach1 and part1.

Create two materials matl1 with capacity 50 and initial amount 25, and matl2 with capacity 100 and initial amount 50.

Create a Demand with the following specification:
Number of Parts to Release : 5
Number of Parts per Load : 2
First Release Time : 0
Inter-arrival Time : normal(.9,.1)
Part : part1

Create the following list of Jobsteps:

Type	Description
Assemble	1 of matl1, 2 of matl2 per Load
Operation	Allocate/Free mach1
	Operation Time = .1
Add-to-Material	1 unit of matl1 per Load
Add-to-Material	2 units of matl2 per Load

Run the simulation.

7.2.4 Accumulation and Splitting

Accumulation is the process whereby separate Parts from the *same* Order are aggregated into larger Loads and are not processed until the larger aggregation is available. If however the total number of Parts in the Order have arrived at the Jobstep the accumulation quantity is ignored.

The reverse process is **Splitting** which disaggregates Loads or Parts by creating new Loads or Parts. The **Accumulate/Split Jobstep** is a refinement of the Operation Jobstep which deals with cases in which either accumulation occurs before the operation proper or splitting occurs after it, or both. The sequence within the Jobstep is always Accumulate - Operate - Split.

As an example Parts may be palletized before an Operation is performed on them (accumulation) and subsequently be de-palletized. Indeed any situation in which Parts arrive from a work cell and are counted into aggregates of size n before the work cell commences its operation, and then despatched in possibly different aggregations of size m, can be described by the Accumulate/Split Jobstep. As with the Release Jobstep the question arises of what to do about imbalances between m and n, and so Rules must be made concerning Excess Options.

The following small Alternative illustrates the basic operations of the Accumulate/Split Jobstep.

Create mach1 and part1.

Create a Demand with the following specification:

Number of Parts to Release : 4

Number of Parts per Load : 4

First Release Time : 0

Inter-arrival Time : .1

Create var1 and a Time-Persistent Value to display the count of Parts after mach1.

Create a Process Plan with two Jobsteps:

Type	Description
Accumulate/Split	Basis = Parts
	Accumulate = 12
	Split = 3
	Allocate/Free mach1
	Processing time = .15
Assign	var1

Run this for two hours so that 20 Orders are generated.

View the Alternative Summary.

Since the number of Parts in each Order is four, which is less than the Accumulation quantity, the latter is ignored. Also, because the processing time is less than the inter-arrival time for Orders, mach1 is continuously busy and 13 Orders are completed at a Production rate of 26 Parts per hour.

The Observed Statistics for Makespan are

$$(.45, .195, .15, .75)$$

(recall from section 1.6 that the order of these is average, standard deviation, minimum, maximum, count if appropriate) showing how the congestion at mach1 is building up. This is also confirmed by the Observed Statistics for Waiting Time which are

$$(.30, .195, 0, .6)$$

and by the average numbers of Loads processing, waiting and in system which are 1, 3.325 and 4.325.

Now introduce two further Jobsteps which will force accumulation to become operative.

In dord1 change the Number of Parts to Release to 12.

Insert at the start of pplan1 the following two Jobsteps:

Type	Description
Probabilistic Select	see below
Operation	**Operation time** = normal(.2,.1)

In the Probabilistic Select Jobstep give a 0.5 probability that each incoming Load is directed to the Accumulate/Split Jobstep and the same for the Operation Jobstep. There is no need to associate a Resource with the Operation - this Jobstep is simply used to delay a proportion of the Parts which make up a Load.

Select a Dynamic Graph for the state of mach1.

Run this for two hours so that again 20 Orders are generated.

The Dynamic Graphs, the Alternative Summary and the graphs after the run should all confirm the delay experienced on account of the requirement for an accumulation of 12 before mach1 processing can commence.

7.2.5 Batching

Whereas accumulation can be performed only on Parts belonging to the *same* Order, **batching** can be performed on items which originate in *different* Orders. This describes the situation where for example different Parts in different Loads must be kept separate for assembly purposes but may be brought together in a wash Operation. Batching can also be used to re-aggregate Parts into Batch Loads which are homogeneous with regard to either Family, Subfamily or Part-number. A Batch is a non-graphical entity in its own right, like a Breakdown or a Shift.

The following Alternative models a situation where two different Part types are batched in the first step of a Process Plan. They then take part in a common Operation (say a wash process), and then a Select Jobstep separates them again into two different Operations depending on their original Part types.

Batching can be either **local**, that is, when batching and unbatching take place within the same Process Plan as in the example below, or **remote**, when batched Loads proceed to a different Process Plan for their batch processing, and then return to their original Process Plan following unbatching.

Create part1, part2, mach1, mach2 and mach3.

Create a Batch batch1 with min/max release quantities = 10 and 15 and Rule = **One per Load.**

Create var1 and var2 and also two Time Persistent Values to display them alongside mach2 and mach3 respectively.

Create two Demands with the following specifications:
Number of Parts to Release : 5
Number of Parts per Load : 2
First Release Time : 0
Inter-arrival Time : normal(3,.5)
Part : part1

Number of Parts to Release : 4
Number of Parts per Load : 4
First Release Time : .2
Inter-arrival Time : normal(.4,.1)
Part : part2

Create the following list of Jobsteps:

Type	Description
Batch	**Name** = batch1 ; check **Local**
Operation	**Allocate/Free** mach1
	Operation Time = exponential(.1)

Select	Rule = Condition
Assign	add 1 to var1
	Allocate/Free mach2
	Operation Time = .1
Assign	add 1 to var2
	Allocate/Free mach3
	Operation Time = .1

In the Batch Jobstep define the Operation Jobstep as the **Unbatch Jobstep**.

Complete the Select Jobstep by providing the two conditions partname = = "part1" and partname = = "part2" as **Expr's**. This directs flow to one or other of the two Assign Jobsteps. Take care to use both double equals sign and double quotes, and also to make sure that neither Assign Jobstep has a Next Jobstep.

Run the simulation.

To summarize, the characteristics of those Jobsteps which deal with changing aggregate sizes are:

Jobstep	applies to ...
Change-Load-Size	Loads only
Produce	Materials, Loads
Assemble	Loads or Parts, Materials involved
Accumulate/Split	Loads or Parts, all within the same Order
Batch	Loads only, Orders can be mixed
Release	Loads via Orders

7.3 Movement

The second source of complexity arises on account of the need for Parts and Loads to *move* around a factory. When moving vehicles and Conveyors start to move asynchronously congestion becomes a possibility, and there is thus the need for *control*. First the number of essentially different generic types of Material Handling Device have to be identified, and then the control problems investigated, some of which are common to all types, and others peculiar to the individual types.

7.3.1 Material Handling Devices

All factories have a requirement for Parts to be moved at some stage in their manufacture, and the term Material Handling Device is a well-established phrase which covers the range of possible mechanisms for moving Loads. It has to be emphasised that such devices may, in the terminology of this book, carry Parts, Loads and Batches, but *not* Materials since Materials are a form of inventory whose levels are changed in the course of Add-to-Material, Remove-from-Material, Produce and Assemble Jobsteps.

There are three categories of Material Handling Device with recognisably different characteristics. These are

Transporters;
Conveyors;
AGVs.

This chapter starts by discussing Segments and Control Points. These are features common to all of three types of Material Handling Device. It goes on to discuss in turn those features which are peculiar to each category.

7.3.2 Systems, Segments and Control Points

All three types of Material Handling Device operate within **Segments** which in the terminology of Graph Theory are arcs which meet at nodes which are called **Control Points**. Segments are bi-directional, that is, travel is possible in both directions along them. Each Segment has two Control Points, one at each end. Control Points are the only points in a Segment at which routing decisions may be made. A collection of Material Handling Devices all of one type, together with their associated Segments and Control Points, is called a **Transport System**.

Each Control Point has a **Next Segment** defined and may also have an **Alternate Segment**. **Segment Rules** can be specified to deal with situations of blockage or non-availability of space. A selection of these is:

1. Always use Next Segment;
2. If Next Segment blocked or stopped:
   use Alternate Segment
   use Alternate Segment but only if it is not blocked or stopped;
3. If no space on Next Segment:
   use Alternate Segment
   use Alternate Segment but only if it is has free space.

Routing Rules take a broader view of the entire path through the material handling system. They presume that paths for each trip are calculated dynamically, and specify that the Next Segment be chosen on the basis of:

Shortest Path;
Shortest Available (that is, unblocked) Path.

Two Segments may share a Control Point, so that a network of Segments is an undirected graph with Control Points as nodes. Two special types of Control Point are called **Pickup Points** and **Dropoff Points**. Since more than one Segment can converge on a Control Point, the latter are potential sources of congestion and contention. This aspect of Material Handling Devices is discussed in section 7.4.5.

The principle of Segments is illustrated by the following fragment.

Invoke the Second Shot Alternative (see section 4.8).

Add a second Conveyor Segment which shares its Control Points with the existing one. This requires some care in positioning the mouse before clicking.

Ensure that the second segment has a visibly different path from the first, for example by leading it round three sides of a rectangle of which **cseg1** forms the fourth. Single clicks cause the path to change direction but do not terminate the Segment.

In the Facility Window click on **ccp1** and ensure that the **Routing Rule** is **Shortest Available Path**.

Create a Breakdown and associate it with a Conveyor (radio button) and cseg1 (list box). Make it breakdown after the first hour with a **Repair Time** 1.

Run the simulation. You should observe the Loads using cseg1 for the first hour, then switching to cseg2 for another hour and finally returning to cseg1.

The same principles of Segments apply to all three types of Material Handling Devices, and so the following sections will focus on the most important differences between them.

7.3.3 Conveyors

Conveyor Systems may be **Segmented**, that is with fixed positions for Loads, or **Continuous**, that is like a conveyor belt in having no fixed positions for Loads. **Spacing** is an attribute of Segmented Conveyors, that is there is a minimum distance (possibly zero) between adjacent Loads on the Conveyor.

Length and **Velocity** are fundamental properties of a Conveyor, and so is **Capacity**(that is, the maximum number of Loads which it can accommodate at any single point in time taking Spacing into consideration). In AIM the length of a Conveyor Segment is always given in feet. Its value may be changed by using the Conveyor editor. The default length is defined by its representation in the Facility Window which in turn is based on the value in the Segment length per pixel field in Build Options.

Following a Breakdown the Loads attempting to use a Conveyor Segment may be **Blocked** or, in the case of a Continuous Conveyor only, **Accumulated**. A Breakdown results in the failure to allocate a Resource at Dropoff. There are three possibilities in this situation:

the Load becomes blocked;
the Load bypasses the broken-down Segment;
the Load waits for a time and then bypasses the broken-down Segment.

Indexed Conveyors are a specialised form of Segmented Conveyor in which each Conveyor position effectively behaves like a miniature Machine. In AIM Indexed Conveyors must be modelled using a Process Plan which contains a sequence of alternating Move and Operation Jobsteps.

127

7.3.4 Transporters

The word **Transporter** is a generic name for any device which can *in reality* travel freely, that is, it is unconstrained by rails or similar guidepaths in its task of conveying Loads from one place to another. Typical examples are fork lift trucks and manually operated trolleys. In simulation models paths and routes for Transporters must be pre-defined, which is the reason for including the phrase "in reality" above. Transporters may have different maximum speeds when **empty** and **loaded**. A collection of Transporters, all of which must possess the same characteristics, is known as a **Fleet**.

A **Transporter System** consists of **Segments** and **Control Points,** and a Fleet may only operate within one Transporter System.

7.3.5 AGVs

An AGV (Automated Guided Vehicle) is a like a Transporter except that it is driverless and *is* constrained to run in specific guidepaths. All vehicles in an AGV Fleet are assumed to be managed by a computer loaded with a control system which contains the logic to determine which particular vehicle is dispatched to a job when more than one is available (vehicle Selection), and also how to sequence Loads for which movement demands are made but no vehicle is available. In AIM the lengths and the number of vehicles in a Fleet may be varied, and also the speeds of the vehicles when empty and full, subject to the assumption that all vehicles in a Fleet are homogeneous. The logic within the computer also deals with collision avoidance, deadlock and contention at guidepath junctions. AGVs are normally constrained to maintain a minimum spacing, typically one vehicle length. Also where Segments intersect a **Check Zone** is defined by radius, and only one

128

vehicle at a time is allowed within this zone. When a vehicle has dropped off its Load there are three possible next actions it may take:

stop where it is;
cruise;
move to some specific location to park.

7.3.6 Transporter and AGV Selection Rules

When a vehicle from a Fleet is required to convey a Load, logic at the level of the Transporter or AGV System must apply a rule to determine which of the available vehicles answers the call. This may be the closest vehicle, or the closest cruising vehicle, or the closest parked vehicle, or the longest idle vehicle.

7.4 Sequencing, Selection and Contention

The third source of complexity springs from the need for objects which are in contention to be *sequenced* in the course of forming queues, and then as a separate issue *selected* for service when this becomes available. In simulation terms this is the area of a simulation system or package which is most algorithmically dense and thus most complex.

7.4.1 Sequencing and Selection

This section is about **contention**, that is, what happens when Resources, Pools and Material are not sufficient to deal with the demands made on them, thus leading to the formation of queues. Logic has to be constructed to deal with the many ways in which such queues can be managed.

Sequencing is about the different ways in which objects in a queue can be *ordered*. Selection is about how items are *removed* from the queue at the moment when the appropriate Resource or Resources become available. Selection is a pervasive and often subconscious activity in manufacturing. A line manager or group leader must react to a congestion situation in which the build-up of incoming material is greater than his or her immediate capacity to process it. Sequencing procedures determine how the Loads are organised in the waiting area. However, when a Resource becomes free, choices have to be about what has to be done first, and the possible bases for making such decisions may not be the same as those which operated in making sequencing decisions.

Many algorithms have been included in AIM which allow the user to select from a rich variety of possible ways of making both sequencing and selection decisions. It has to be said that the succinct terms which are used to describe these options in list boxes can be puzzling, and it is a good idea to sort out in advance and in general terms the options available for controlling queues, whether real ones or simulated ones.

7.4.2 Sequencing Rules

The order in which events are sequenced can be of the highest importance in determining the performance of a manufacturing system. Even quite simple systems can be made to exhibit a wide range of sequencing rule possibilities, and one of the greatest potential benefits of simulation is to allow the comparison of different rules in both existing and planned systems. It sometimes happens that the application of different sequencing rules alone has a dramatic effect on the ability of a system to meet its requirements.

Slack is an important concept in sequencing. There are two sorts of Slack, **Static** and **Dynamic**. Static Slack means Time remaining to Due Date. Dynamic Slack means Time-to-Due-Date minus Processing-Time-Remaining.

A list of the main Sequencing Rule possibilities is given below. They are common to actions within Jobsteps which affect Resources, Pools and Materials and Material Handling Devices. Also each Alternative has a Global Sequencing Rule set by default to FIFO (First In, First Out). This can be changed to any others from the list, and the Global Rule can be over-ridden for individual Resources. The list is:

1. Sequencing by Queue Discipline:
   FIFO (First In, First Out);
   LIFO (Last In, First Out).
2. Sequencing by Integer values:
   Attributes (values assigned to Loads by Assign Jobsteps);
   Priority (integer from 0 to 9999 - high value = high priority);
   Load-size;
   Number of Jobsteps.
3. Sequencing by Date/Time:
   Date (Due or Release);
   Jobstep Time (Max or Min for Current or any subsequent Jobsteps);
   Processing Time Remaining.
4. Sequencing by Least Slack - Static or Dynamic:
   Absolute;

Relative to number of Jobsteps;

Relative to Processing Time Remaining.

7.4.3 Selection Rules

The simplest Selection Rule is to accept sequencing. Other alternatives might be to do deliberate change-overs in order that certain types of Loads, for example, those with Parts belonging to the same Family or Subfamily or possessing the same Part-number, should not be allowed to exclude other Load types. Change-overs of this sort might be made after a specified number of either Loads or processing hours. The criterion for what to select at such a point might be the Setup request which has waited longest, or alternatively the one which has the shortest remaining Setup times in total. Another option is to examine and apply Selection Rules only to the first n outstanding requests as ordered by sequencing, where n is some predetermined integer.

7.4.4 Order Release Rules

This is to Orders what selection is to Loads. There is a variety of possible criteria for selecting which Order to release in a situation where more than one is scheduled for release at the same time. These include:

Earliest Due Date;

Number of Jobsteps;

Order Size;

Processing Time;

Priority;

Least Static Slack (that is, time to due date);

Least Static Slack divided by number of Jobsteps;

Least Static Slack divided by processing time.

To reset the Global values for the Sequencing and Order Release Rules in AIM do the following:

Alternative action bar **Options ; Simulate..**

Reset Global Sequencing and Order Release Rule as required.

7.4.5 Contention Rules

Material Handling Devices also use their own selection rules when several Loads are waiting to enter a Control Point. The criteria involved in these rules are:

Earliest Due Date;

Earliest Order Release Date;

FIFO or LIFO;

Attributes;

Priority;

Largest or Smallest Load size.

7.4.6 Control Point Groups

Machines within a Resource Group may be separated by distances which are significant as far as transport movements are concerned. It is convenient to define a **Control Point Group** whose members are in one-to-one correspondence with the members of a Resource Group. The Control Point Group may then be used either as an origin or as a destination in a Move-Between Jobstep, and when one member of the Resource Group is selected for *allocation*, the corresponding member of the Control Point Group is selected as the *location* for vehicle movement.

8. EXPERIMENTATION

8.1 Client/User Relationship

The word "user" has been used so far with the implication that the person who inputs models, performs simulation runs and interprets the results is the same person who was motivated to set up a simulation project in the first place. With simulation work of any substance this situation is very unlikely, and it is much more common in practice to find that there is on the one hand a client who first proposed (and possibly funded) the general idea that something, e.g. line design or operation, should be investigated using simulation, and on the other hand a user who possessed the computing and modelling skills to transform that idea into a reality. Of course either or both of these parties may be plural, but for the sake of this discussion they will be referred to in the singular.

No matter how great the user's modelling skills, they cannot be put to good use unless they are deployed at the right time. In the case of design, where the objects being simulated do not yet exist, this means that with sufficiently early involvement changes can be made in response to the evidence of simulation. Where simulation is conducted as a means of trying to improve operations on existing equipment, or to try out radical alternatives whose expense could not be contemplated in terms of the disturbance to routine production, then again it is important that the user should not only capitalize on the client's enthusiasm, but also make very deliberate efforts both to involve the people most directly concerned and to persuade them that in some sense the model belongs to them, since without their input data it would not exist.

The user should also be prepared to accept the possibility of sudden loss of interest on the part of the client. This is no new phenomenon. It happens in the moment of the *client's* insight, when what is important to him or her suddenly becomes blindingly clear. At this point the user has probably become deeply involved with the model. He or she may well be planning to refine and enhance it, and to use it as a test-bed for the sort of hypothesis testing and experimentation which is the foundation of true science.

This preamble is given to record the practical reality of the relation between simulation and experimentation. In particular, attempts to use designed experiments in the sense that statisticians understand this term are often stillborn. The user should be prepared not to feel disappointment in this situation. The simulation project terminates when the client either becomes persuaded that his objective has been achieved or decides to channel his urge for scientific computing in some other direction, or more probably gets caught up in the exigencies of day to day crisis management. (Management science is all right when you are not in a hurry!) The user cannot reasonably foresee when such a moment will arrive.

Several books which discuss simulation contain material on experimentation as if it was a natural and inevitable step in a smooth and ongoing process. In fact it is a step which, for the reasons just given, may never be reached, and so what follows is in some ways of a counsel of perfection.

Simulation today is very much bound up with visual interactivity, and the user would be foolish not to court the client by taking extra time and trouble on those cosmetics which make the visible outcome of his or her work as pleasing and effective as his computer and his computing skills allow. An ABCD of such cosmetic features is:

Annotated Output, that is, traces of individual runs or sections of runs which persuade the client that the processes in the computer accurately mirror on a step by step basis the analogous actions in the real world;

Business Graphics, charts, particularly comparative charts, of quantities such as those listed in section 4.2.3;

Conversational Menus offering the possibility to change parameters easily for each new run; and

Dynamic Animation - perhaps the most powerful weapon of all in providing visual evidence that the computer is acting as a mimic of some actual or planned reality.

This book so far has been concerned with model development, that is that phase in which the user understands the situation posed for him by the client, and goes about the construction of Problem Configurations, Databases and Alternatives. This stage often has merit in its own right in clarifying the understanding of both client and user regardless of whether any

137

explicit numerical results at all are produced. The discipline forced by a computer package on the user results in his or her insistence in extracting answers to questions which the client may have been, albeit subconsciously, shirking. In this role the activities leading up to simulation, that is, requirements analysis and model design, can often be as valuable as the simulation runs themselves.

8.2 Designed Experiments

Although models which proceed no further than requirements analysis and design may have some usefulness, a project which gets no further than the stage of a few runs is probably not realising its full potential. The next stage of experimentation is reached when a model is agreed to be correct by both user and client - the processes taken to reach this stage are collectively called **validation** - and correct Alternatives are run repeatedly with different parameter combinations with the object of either predicting the effect of anticipated future changes in a manufacturing complex, or of optimizing those parameters on which its overall performance depends. This is a further point at which a simulation package or system must be complemented by other skills and software which cover that part of statistics which is called Design of Experiments. Usually this means no more than applying practical (that is, statistical!) common sense by encompassing repeated runs within logical frameworks such as Factorial and Balanced Block Designs and then following this up with the arithmetic practices which are known collectively as Analysis of Variance. For those enthused by Japanese methods, Taguchi type of experimentation can be validly applied at this point.

It is not the place of this book to describe this part of statistics at length. Many excellent books exist on this subject such as Ref [9]. This one stands out from the rest by its implicit acknowledgement that the power of computers to do instant calculations means that a textbook in this area can and should contain hundreds of worked examples which display the rich variety of possibilities which Design and Analysis of Experiments and related techniques such as Response Surface Analysis afford.

8.3 Variance reduction and Control Variates

The output values or *responses* from simulation runs as prescribed in the Problem Configuration are estimates of what may be hypothesised to be true but unknown values. Clearly, the tighter the distribution of such responses in many Replicates, the better is the estimate of the quantity concerned in the sense of their having a narrower confidence interval. As a matter of professional competence the user should never in a final report present such estimates without confidence intervals, even although the latter may not have been explicitly requested by the client.

Variance reduction is a term used to describe a wide variety of techniques in Numerical Analysis whose aim is to improve estimates (in the sense of narrowing confidence intervals) without increasing experimental effort. In the vocabulary of statistics this is known as making a process more *efficient*.

One method of variance reduction (use of antithetics) has already been encountered in section 3.5.1. Another is the use of Control Variates. This is an application of Regression Analysis which is another major area of statistics.

The fundamental idea is very straightforward, and is rather like that used in making seasonal adjustments to economic statistics. A full discussion, however, requires the parallel availability of a good standard statistics book which covers such topics.

Suppose that the values of a response variable, y say, are judged to be related to some other variable x by means of a linear relation:

$$y = k_0 + k(\bar{x} - E(x))$$

where k and k_0 are constants and $E(x)$ is the Expected Value (that is, long run average) of the Control Variate x. Least squares theory directs that the value of k which gives the best fit is given by

$$k^2 = \rho^2 \frac{var(y)}{var(x)}$$

where ρ is the correlation coefficient of x and y. This value of k can be used to obtain an adjusted estimate $\hat{y}$ from the observed average $\bar{y}$ in a series of replicates using

$$\hat{y} = \bar{y} - k(\bar{x} - E(x))$$

The variance of $\hat{y}$ is

$$(1 - \rho^2) var(\bar{y})$$

If the correlation between x and y is high enough the variance of $\hat{y}$ will be less than that of $\bar{y}$, and so a more accurate estimate of performance can be given in the sense that a lower confidence interval can be quoted.

The problem with this technique is that the quantities ρ, $var(\bar{y})$ and possibly $var(\bar{x})$ have themselves to be estimated. It may be possible to do so them from historical records; if not, they have to be estimated by doing screening runs prior to the main simulation, which in turn requires more experimentation. It could be that this extra effort might be better deployed in simply increasing the number of replicates in the main experiment and achieving variance reduction from the fact that the variance of $\bar{y}$, that is the mean of a sample of size n, is related to that of the variance of y by the formula

$$var(\bar{y}) = \frac{var(y)}{n}$$

In order that the Control Variate technique be beneficial it is necessary that ρ^2 be at least as great as $1 - \frac{1}{n}$.

The number of Control Variates may exceed one. In this case separate coefficients k_i must be estimated for each variate and the corresponding adjustments added together.

8.3.1 Numerical Example of a Control Variate

Consider a warehouse simulation in which one of the objectives was to measure long-term throughput measured as pallets per shift (y) delivered out of the loading bay area. It was believed that y was related to the proportion of small vans in the mix of incoming delivery vehicles. Call this proportion x. The design point is that x = 0.5, and in the course of modelling, all Alternatives use this value as the basis for random number generation. In any *particular* run the simulated value of x will vary from the Expected Value of 0.5. If the value which is of interest to the client is the throughput perform-

141

ance of the warehouse *relative to x*, or equivalently the performance which would be achieved under a steady state of $x = 0.5$, then the values of y obtained in a simulation run or runs should be adjusted to take account of the values of x.

Suppose that five runs gave the following paired values of x and y:

y	x
161	0.425
164	0.512
173	0.638
153	0.392
161	0.563

The correlation coefficient is 0.87 and the regression equation is:

$$y = 162 + 62.4(x - 0.5)$$

This means that for any subsequent replicate or replicates, instead of reporting the *actual* mean response $\bar{y}$, the reported value is

$$\hat{y} = \bar{y} - 62.4(\bar{x} - 0.5)$$

For example, if the observed value of $\bar{y}$ is 165 and $\bar{x} = 0.58$ then report

$$\hat{y} = 165 - 62.4 \times 0.08 = 160$$

Since $n = 5$, the correlation coefficient exceeds $1 - \frac{1}{n} = 0.8$, and so in this case the use of a Control Variate improved the quality of the information reported.

Confidence intervals for *predictions* of the true value of y depend on the value of x for which the prediction is made. In the present example standard

142

statistical calculations show that variance reduction is indeed achieved by using $\hat{y}$, since the standard error of *prediction* of $\hat{y}$ is around 5, whereas the estimated standard error of the mean for the five values quoted above is about 10.4.

APPENDIX A: Major AIM Components

Graphical

1. Resources
(single and multi-capacity):
 	
    Machines
    Operators
    General Resources
    Fixtures

2. Inventory:
 	
    WIP
    General Pools
    Material

3. Transport - Devices:
 	
    Conveyors
    Transporters
    AGVs

4. Transport - Routes:
 	
    Segments
    Control Points

5. Display: Time Persistent Values

Non-graphical

1. Process Plan

2. Production:
 	
    Part
    Order
    Batch
    Load

3. Order Creation:
 	
    Demand
    Pull

4. Groups:
 	
    Resource
    Control Point

5. Systems:
 	
    Conveyors
    Transporters
    AGVs

6. Interruptions:
 	
    Breakdown
    Maintenance
    Shift

APPENDIX B: AIM Jobstep Types

1. Basic Set	notes

a. Making:

Operation	
Setup	by new Load, Family, Order, etc.
Setup/Operation	combines both of the above
Kanban	Operations controlled by Tags

b. Moving:

Move	no resource involved
Move-Between	Material Handling Device involved

c. Inventory:

Add-from-Material	often last Jobstep in Process Plan
Remove-from-Material	often one of first Jobsteps in PP

d. Display:

Assign	to Variable or Attribute

2. Branching:

Probabilistic Select	by prob/Jobstep pairs, $\sum p_i = 1$
Select	by condition, list, or cycle
Inspect	Operation followed by Selection

3. Aggregation/disaggregation:

Change-Load-Size	Loads only
Produce	Materials, Loads
Assemble	Loads or Parts, Materials involved
Accumulate/Split	Loads or Parts in the same Order
Batch	Loads only, Orders can be mixed
Release	Loads via Orders

APPENDIX C: Two Further Concepts

A summary of AIM is also a general summary of manufacturing concepts. Two further concepts which do not occur explicitly in the AIM vocabulary are described below with pointers to how they might be modelled.

1. Lead Time	This is the time delay which elapses between a reorder of Material, and its subsequent receipt. When an Add-to-Material Jobstep is triggered by a Pull this delay can be modelled by: (a) selecting Need as the Pull Release Rule (see secn. 5.1); (b) including a Remove-from-Material Jobstep for the number of units expected to be used in the course of the Lead Time; and (c) issuing an Add-to-Material as the first Jobstep in a Process Plan triggered by the Pull. The number of units added may be, but need not be, the same as that removed in (b). The numbers of units in (b) and (c), together with the Material capacity limit, provide fine tuning possibilities.
2. Expedition	As a general manufacturing principle, Orders and Loads should be processed in a disciplined fashion without delays due to rogue Loads or Orders lingering for long times in process. The modelling controls available for this are Priority (secn. 7.4.2) and Integrity (secn. 2.2.5). Priority means assigning an integer value to an Order, and thereby to Loads within the Order, which influences Selection (secn. 7.4) at Resource Allocation. Priority is by default in ascending numeric order, and values may be changed dynamically using the Assign Jobstep. Integrity, on the other hand, is the principle of keeping Loads and Orders together, i.e. of allocating Resources in such a way that an individual Load or Order is completed before a new one is embarked on. Gathering of this sort may take place either at a single Jobstep, or universally. It is part of Setup preceding an Operation, and requires that the first Resource allocated in each affected Jobstep is single-capacity, and has a non-zero Setup time.

REFERENCES

1. Allen, A.O., 1978, Probability, Statistics and Queueing Theory with Computer Science Applications, Academic Press

2. Banks, J. and Carson, J.S., 1984, Discrete Event Simulation, Prentice Hall

3. Davies, R.M. and O'Keefe, R., 1989, Simulation Modelling with Pascal, Prentice Hall

4. Evans, M., Hastings, N.A.J., and Peacock B., 1993, Statistical Distributions, 2nd. edition, John Wiley and Sons

5. FACTOR/AIM Users Guide and Modelling Reference, 1993, Pritsker Corporation, 8910 Purdue Road, Indianapolis, IN 46268

6. Fishman, G.S., 1978, Concepts and Methods in Discrete Event Digital Simulation, John Wiley and Sons

7. Law, A.M. and Kelton, D., 1982, Simulation Modelling and Analysis, McGraw-Hill

8. Lewis, C.D., 1970, Scientific Inventory Control, Butterworths

9. Montgomery, D.C., 1991, Design and Analysis of Experiments, 3rd. edition, John Wiley and Sons

10. Page, E., 1972, Queueing Theory in OR, Butterworths

147

REFERENCES

Index

149